Atomic Layer Deposition

Scrivener Publishing
100 Cummings Center, Suite 541J
Beverly, MA 01915-6106

Publishers at Scrivener
Martin Scrivener (martin@scrivenerpublishing.com)
Phillip Carmical (pcarmical@scrivenerpublishing.com)

Atomic Layer Deposition

Principles, Characteristics, and Nanotechnology Applications

Tommi Kääriäinen, David Cameron, Marja-Leena Kääriäinen, and Arthur Sherman

2nd Edition

Scrivener Publishing
WILEY

Co-published by John Wiley & Sons, Inc. Hoboken, New Jersey, and Scrivener Publishing LLC, Salem, Massachusetts.
Published simultaneously in Canada.

For general information on our other products and services or for technical support, please contact our Customer Care Department within the United States at (800) 762-2974, outside the United States at (317) 572-3993 or fax (317) 572-4002.

Wiley also publishes its books in a variety of electronic formats. Some content that appears in print may not be available in electronic formats. For more information about Wiley products, visit our web site at www.wiley.com.

For more information about Scrivener products please visit www.scrivenerpublishing.com.

Cover design by Russell Richardson

Library of Congress Cataloging-in-Publication Data:

ISBN 978-1-118-06277-7

10 9 8 7 6 5 4 3 2 1

Contents

Acknowledgements

Tommi and Marja-Leena Kääriäinen are grateful for their children Emil, Elsa, and Elias and would like to thank for their love and inspiring moments in life.

David Cameron dedicates this work to his wife, Eileen, for her unfailing love and support and to his children Petra and Alasdair and granddaughter, Rose.

Arthur Sherman would like to acknowledge the support of his family (daughters Linda and Jan, and son Douglas, as well as four beautiful grandchildren Ashley, Daniil, Andrew and Sara). An affectionate thank you is also extended to Sandra Tenenbaum

Foreword

The field of atomic layer deposition (ALD) has experienced tremendous growth over the past ten years. The Web of Science reveals that nearly 900 publications in 2012 and over 5000 publications since 2002 had "atomic layer deposition" in their abstracts or key words. If we include the earlier use of "atomic layer epitaxy" to refer to the field, then we can add over 1500 more publications from the past thirty years. Why has the growth of ALD increased so dramatically over the past ten years? There are many contributing factors that include the development of new ALD materials and the availability of commercial ALD reactors. The wealth of ALD materials means that many applications can be addressed with known ALD chemistries to deposit the thin film. The many companies selling ALD reactors means that ALD practitioners can concentrate on their applications and do not need to worry about building their own ALD reactors.

Another important factor to explain the dramatic growth of ALD over the past ten years is that more people are becoming aware of the superior features of ALD and the importance of these features to nanofabrication technology. These characteristics have been well known since their introduction by Suntola in the early 1980s. The virtue of sequential, self-limiting surface chemistry to achieve conformal and atomic layer controlled growth was presented over thirty years ago. Although many important ideas are recorded in the literature, only a few really come to fruition and become the basis for an entire industry. One key event that stimulated interest in ALD and subsequent recognition for ALD was the semiconductor industry realizing that they needed ALD for high K gate oxides to replace SiO_2. This pivotal event occurred approximately 10-12 years ago when ALD was placed prominently on the semiconductor roadmap for high K gate oxides.

This revised book on ALD is a welcome addition to the field of ALD and will be very useful for experts and beginners alike. The book provides an up-to-date review of ALD including the rapidly developing plasma ALD and spatial ALD processes. The core of the book provides a description of the variety of ALD chemistries for semiconductor, oxide, nitride, metal and hybrid material films. Many of these ALD chemistries did not even exist ten years ago. One chapter is dedicated to semiconductor epitaxial films and reviews the early work on silicon and germanium ALD. Another chapter describes the important oxide ALD films that have pioneered many of the important applications for ALD over the last ten years. The book ends with a summary of the current ALD applications that are driving the field and leading to new ALD commercialization. Most of these applications were only on the drawing board ten years ago.

This book should also serve as a springboard for future developments of ALD. Knowledge of the past is an important foundation for moving into the future. As the ALD field moves forward, we will want to identify: the key challenges that are currently limiting progress in ALD; the critical developments that are needed to advance ALD state-of-the-art; and the new applications that may benefit from ALD methods. This book provides the present ALD status quo that will provide a baseline to measure future progress. The last ten years have brought tremendous progress, and continued growth is on the horizon. One thing is for certain: ALD will continue to be an integral part of nanofabrication technology as we move into the future.

Steven George
March 2013

Preface

Preface to First Edition

More than 25 years ago, researchers in Finland realized that films only nanometers thick could be deposited uniformly by chemical reactions on surfaces, by a novel new process. This process made use of the fact that, in general, when a substrate within a vacuum chamber is exposed to a reactive gas, that substrate will retain a single monolayer of that gas even when the gas is evacuated from the chamber. When this monolayer is exposed to a second reactive gas, causing a chemical reaction whose products include a solid species, a monolayer of this solid is formed. Repeating this process allows films to be grown one monolayer at a time, and is referred to as Atomic Layer Deposition or ALD. Over the years the ability of this process to deposit thin films conformally, with extraordinary thickness control, and uniformly over substantial areas, has led to an increasing level of research and development. In fact, the number of published scientific articles on ALD has more than doubled over the last five years.

Since this new technology is quite interdisciplinary, a review of it from the point of view of a single author would be very useful. That is the motivation that has led me to put this text together. Although an attempt has been made to cover most important films and processes, it definitely has not been my intention to draft an encyclopedic text. Just enough material has been included as was felt necessary to convey the important concepts underlying ALD.

Arthur Sherman
January 2008

Preface to Second Edition

Since publication of the first edition of this book, research into atomic layer deposition has continued to expand; the annual number of publications explicitly mentioning atomic layer deposition, ALD or molecular layer deposition as a topic has grown by two thirds between 2008 and 2012 and many more research groups are becoming involved in this field. In addition, it has become well established for many industrial applications. The long-established commercial application of ALD to electroluminescent displays has been followed by integration into, for example, microelectronics and magnetic recording head production. In addition, other new areas such as protective coatings for jewelry have grown up, and the application of ALD to many diverse end products is being pursued vigorously.

There have also been numerous developments in the ALD process itself. The number of materials which have been produced is now very large and continues to grow. New precursor chemicals are being developed to improve their ease of handling, reduce their temperature requirements and facilitate the chemistry of the process. The use of reactive-enhanced ALD has seen continued improvement and the availability of spatial ALD making possible roll-to-roll systems is an exciting development. At the same time, deposition of organic and hybrid organic-inorganic layers is now a popular area which promises interesting and useful capabilities. Finally, the availability of well-engineered ALD systems has enabled turnkey operation, and has reduced the hurdles for entry to the technology.

In the light of these developments, there are many people for whom an updated text dealing with the principles, technology and applications of ALD would be valuable. With this in mind, the structure of the book has been somewhat altered. Chapter 1 still deals with the fundamentals of the process, but greater emphasis has been put on reactive-enhanced ALD and the principles of spatial ALD are now outlined. There has been limited development in the epitaxial growth of elemental and compound semiconductors, therefore Chapters 2 and 3 are relatively unchanged. Chapter 4 on oxide coatings has been extensively updated to take account of recent work and new precursors. Reactive-enhanced ALD processes have been integrated into this and other chapters rather than occupying a separate chapter. An expanded chapter 5 deals with nitrides and other compound materials. Metal film deposition has greatly developed in recent years and Chapter 6 is devoted to this alone. Chapter 7 deals with organic and hybrid layers, which is a rapidly expanding area. To conclude, Chapter 8 describes applications and examples of industrial use together with a list of some ALD equipment suppliers.

We are aware that the field of ALD is rapidly developing, and that many new applications and process developments will occur. The aim of this text is to introduce the reader to the principles, characteristics, capabilities and possibilities of ALD in an edition which also provides a guide to where more detailed information can be obtained. We trust that our readers will find that we have achieved this goal.

Tommi Kääriäinen
David C Cameron
Marja-Leena Kääriäinen
Arthur Sherman
March 2013

1

Fundamentals of Atomic Layer Deposition

1.1 Chemical Vapour Deposition

Atomic Layer Deposition (ALD) is a more recent variation on the older technology referred to as Chemical Vapor Deposition (CVD) [1]. In CVD, which was originally developed in the 1920s, a mixture of gases flows over a heated substrate causing a thin solid film to grow on the surface. This heated surface has to be hot enough to allow the surface reaction to proceed rapidly, so that commercially acceptable deposition rates are achieved. In the ideal case, there will be no reaction between the reactant gases in the gas phase, in other words no homogeneous reactions, which would cause the formation of particulates. The gases approaching the heated surface will be heated by gas phase conduction and should not react until they impinge on the surface where they form a solid film of deposited material by a heterogenoeous reaction. If a homogenoeous reaction takes place, in the worst case one can have particles forming in the gas phase and ending up embedded in the growing thin film, clearly an unacceptable result.

The optimum choice of reactants for a CVD process is generally a mixture of the most reactive gases available. This allows film deposition at the highest rates, and at the lowest substrate temperatures. Unfortunately, this choice leads to a high probability of gas phase reactions, which as noted can compromise the deposited film quality. It is this dichotomy which has fueled much of the CVD research over the last thirty years.

The present book will be a review of CVD's newer variant, ALD. In this technique, the presentation of the two reactants to the heated surface is separated into two steps. In step one, the substrate is exposed to the first reactant after which this reactant is pumped away. During this exposure a monolayer of the first reactant adsorbs to the substrate, and remains after the chamber is evacuated. Then a second reactant is introduced into the chamber, and it reacts with the monolayer of the first reactant. This then forms one layer (generally less than one complete monolayer) of the solid film being sought. After this, the remaining second reactant and any gas phase reaction products are removed from the chamber. This process is repeated as many times as necessary to grow a film of the desired thickness. The conformality of the film is also excellent, since film growth depends only on the formation of monolayers on the surface and not on the arrival rate of reactants. The time necessary to form such a layer can be increased if necessary for a complex-shaped substrate and the arrival of excess reactant is not important. Clearly with this process, gas phase reactions should not occur, so that one is free to choose the most reactive reactants available and film deposition temperatures can be lower. However, the one disadvantage is that the film deposition rate may be slow. For those applications where thin, uniform and highly conformal films are of interest, this becomes less of a limitation and ALD is an important process.

To set the scene for a description of ALD, the characteristics of the CVD process will be described.

1.1.1 Thermal CVD

The first CVD processes that had major commercial importance were developed when researchers were required to find ways to deposit thin films of dielectrics and semiconductors on semiconductor wafers. These initial processes were all mostly carried out by simply heating a substrate and placing it in a gas stream of one or more reactants. Initially all the processes were run at atmospheric pressure.

One of the first thin films to be deposited was silicon, because in the early days single crystal wafers were not pure enough to allow integrated circuits to be built into them directly. The CVD silicon films, on the other hand, could be grown at very high purity from carefully purified reactant gases. Among the reaction schemes used were

$$SiH_4\,(g) \rightarrow Si\,(s) + 2H_2 \tag{1.1}$$

$$SiCl_4\,(g) + 2H_2 \rightarrow Si\,(s) + 4HCl \tag{1.2}$$

$$SiH_2Cl_2\,(g) \rightarrow Si\,(s) + 2HCl \tag{1.3}$$

$$SiHCl_3\,(g) + H_2 \rightarrow Si\,(s) + 3HCl \tag{1.4}$$

Any of these four reactions can be used to deposit silicon, and the film morphology can be single crystal, polycrystalline or amorphous, depending on the deposition temperature. A typical plot of deposition rate versus deposition temperature for these four reactions is shown in Figure 1.1 below.

As can be seen, the lowest growth rates at a given temperature are for the silicon tetrachloride reaction (Eq. 1.2), because this molecule

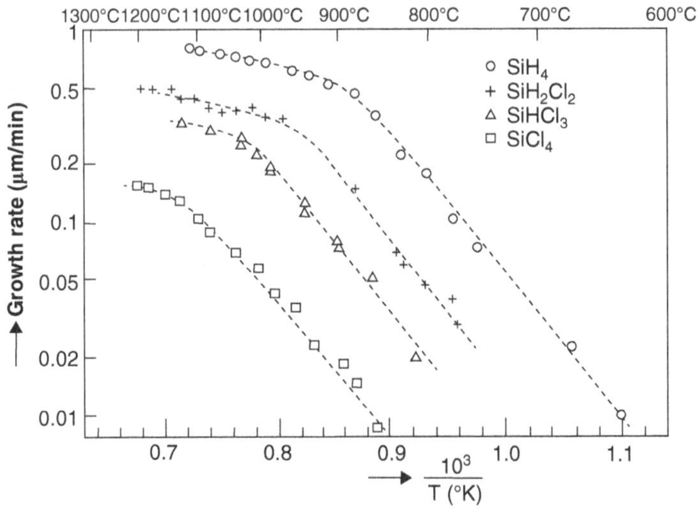

Figure 1.1 Silicon growth rate as a function of temperature [2].

is the most stable of the four silicon precursors. The highest growth rates are seen for the silane reaction (Eq. 1), as this molecule is the least stable of the four. At the higher temperatures, to the left in the Figure, single crystal (epitaxial films) can be grown. At the lower temperatures, to the right, one obtains either amorphous or polycrystalline films.

Generally, the silicon tetrachloride reaction has been commercially preferred, in spite of the high temperatures needed. Partly, this is because this reaction involves molecules that are likely to remain stable as the gas approaches the hot wafer surface, and gas phase reactions are less likely. At the other extreme we have the silane reaction, where this molecule is not a very stable species. Although deposition can be carried out at lower temperatures, there is the risk of gas phase nucleation leading to lower quality epitaxial films.

We also have to note the shape of these curves. When commercially desirable deposition rates are demanded, then deposition at high temperatures is necessary, and we see that the deposition rate is relatively independent of temperature. This is referred to as the diffusion limited regime of CVD. Here, the deposition rate is controlled by the arrival rate of reactant species which diffuse to the surface from the gas stream. The difficulty here is that the film uniformity over large areas depends very much on fluid flow uniformity. Particularly for reactors that coat several substrates at one time, such flow uniformity can be hard to achieve.

If depositions are carried out at lower temperatures, we see that deposition rates vary almost linearly on this semilog plot., that is the deposition rate reduces exponentially with the reciprocal of the temperature. Also we see that deposition rates become very low, and in order for a reactor to be commercially viable one must process many wafers at one time. Now, the deposited film uniformity depends strongly on temperature so it is mandatory that the batch reactor have a very uniform temperature.

Aside from semiconductors such as silicon, many commercially important materials can be deposited as thin films by CVD, such as metals (e.g. tungsten, copper), refractory metal silicides (e.g. tungsten silicide), nitrides (e.g. titanium nitride, silicon nitride), and of course oxides (e.g. silicon dioxide). In addition, CVD is an important process for deposition in non-semiconductor applications such as wear-resistant coating for tools [3] and catalysts [4] and recently for the deposition of carbon nanotubes [5], graphene layers [6], polymers [7] and many other materials.

In summary then, thermal CVD has been a powerful process for the commercial deposition of high quality thin films of many different materials. The process, however, has only proven useful with a restricted range of precursors because of the conflict between the use of the most reactive species in order to obtain the highest deposition rate at low temperature, which may lead to homogeneous reactions and poorer film properties, and the very high temperatures required when less reactive species are chosen but which may give better film properties. One way out of this dilemma is to use a less reactive precursor, but enhance its reactivity by creating a glow discharge in the flowing gases. This topic will be reviewed in the next section.

1.1.2 Plasma Enhanced CVD (PECVD)

Thermal CVD is a process that, although occurring in a flowing chemically reacting gas which is inherently not in thermodynamic equilibrium, occurs close to equilibrium. On the other hand, when a glow discharge is created in a gas at low pressure the resulting plasma is very far from thermal and chemical equilibrium. That is, the electrons in the plasma tend to be very energetic and the ions and molecular species tend to have low energies. For typical situations the electrons can have energies consistent with temperatures of many thousands of degrees, and as a result readily dissociate a substantial number of the molecular species in the reacting gas mixture.

Historically, glow discharges in reacting gas mixtures have been used to lower the required deposition temperatures below those required by thermal processes. This allowed deposition at practical deposition rates of a number of commercially important films. For example, silicon nitride deposited at over 800°C could not be used to passivate an integrated circuit, because aluminum conducting layers would have melted. When silane and ammonia mixtures were used in a low pressure glow discharge, deposition at acceptable rates could be achieved at much lower temperatures ~ 350°C. Accordingly, PECVD of silicon nitride became a standard final passivation layer, and integrated circuits could be reliably packaged in low cost plastic. Prior to this all circuits had to be hermetically sealed in ceramic packages.

There are a number of issues to consider when plasma is used to activate a CVD reaction. The plasma can provide the energy for the chemical reaction and it can create reactive radicals which react more easily than the precursor molecules. However, it can also contain ions and hot electrons which can impinge on the substrate and can have both helpful and deleterious effects on the deposited film and the substrate surface. Bombardment by ions can cause stress, typically compressive stress, which can densify the film and reduce the likelihood of cracking. The ion bombardment can also lead to charging effects in dielectric films which may lead to electrical breakdown. The method of applying the electrical energy to the plasma is also important. The frequency of the activating electromagnetic field can greatly influence the properties of the plasma and its effects. Operating at higher frequencies allows the plasma density to be increased while keeping the electron temperature low and giving lower energy bombarding ions. Operating at frequencies of 80 MHz or higher can greatly increase the deposition rate and combinations of high frequency plasma excitation by VHF or microwave power with a lower frequency substrate bias can decouple the process of plasma generation from that of control of the ion bombardment energy. One effect of increasing the frequency is that the plasma uniformity can suffer when the size of the electrode reaches one quarter wavelength. Standing wave patterns can be set up on the electrode which locally change the plasma density.

As mentioned, plasma enhancement creates an unusual mixture of free radicals that would not exist in the thermal process. In some instances, this feature produces beneficial results, but in other cases the films turn out to have undesirable features. Among the limitations sometimes found are:

1. It is not always easy to control film chemistry.
2. Film density may be low.
3. Film conformality is often poor.
4. Plasma uniformity, and therefore film uniformity, can be hard to achieve.

1.2 Vapour Adsorption

In contrast to CVD, ALD is a two step process where in the first step one reactant has to form a self limiting monolayer on the surface upon which the film is to be grown. What is meant by "self limiting" is that once a complete monolayer of reactant molecules bonds to the surface molecules then the bonding of any other reactant molecules to the reactant molecules that constitute the surface monolayer is unlikely to occur, and one ends up with just one layer of the reactant molecular species. In general, any solid surface exposed to a gas or vapor will retain at least some of the gas or vapor on the surface after all gas phase molecules are removed. The forces holding the molecular species on the surface can be weak (physical adsorption or physisorption), or they can be quite strong (chemisorption). In the latter case, in particular, the adsorbed molecules tend to readily form the self-limiting monolayers of interest for ALD, and such layers can be fairly stable at moderate temperatures. A physisorbed monolayer, on the other hand may be readily desorbed if the substrate surface runs too hot, because of the weak bonding forces. Once a monolayer of a first reactant is formed then the introduction of a second reactant simply converts the first reactant to a layer of some desired solid material. In the current section, we will review some of the salient features of the processes whereby gaseous materials are adsorbed onto solid surfaces.

1.2.1 Physisorption

There are two aspects of adsorption that are essential to the understanding of ALD. First, the bond strength that binds the gaseous species to the surface varies depending on both the gaseous species and the nature of the surface. Secondly, the rate at which adsorption occurs will be finite, and in some cases may be quite slow.

Physical adsorption is characterized by weak bonds to the solid surface (e.g. heats of adsorption < 20kJ/mole), on the order of van der Waals forces, and rates of adsorption that are fast. A "thermal" ALD process is defined as one where the desired surface reaction occurs when both reactants are on the surface at the same time, but will not react until the surface is heated to a high enough temperature. In all "thermal" ALD processes, temperatures, although lower than in CVD, are high enough to cause weakly bonded monolayers to desorb from the surface. For this reason, "thermal" ALD processes generally take advantage of chemisorbed monolayers of the first reactant. However, as will be discussed later in Section 1.3.2, it is also possible to carry out ALD processes with highly reactive free radicals, as the second reactant, at essentially room temperature. In this case, physisorbed first reactants will, in fact, be of interest.

All gases will physisorb a self-limiting monolayer under the correct temperature and pressure conditions, even if they do not chemisorb. The preferred conditions are low pressure and moderate temperatures. For a given pressure, if the temperature is too low more than one monolayer will adsorb, and at low enough temperatures a liquid or solid film will be formed. As the temperature is raised a single monolayer will be left behind on the surface. Finally, if the temperature is raised enough the single monolayer will be desorbed. In general, a single monolayer of a gas will adsorb to a smooth surface as long as, for the temperature being used, the pressure stays well below the saturation vapor pressure [8]. However, when molecules from the gas phase impinge on a surface and begin to stick, a complete (100%) monolayer may not be formed, depending on the pressure or concentration in the gas phase. The simple Langmuir isotherm, illustrates this effect [9]. Langmuir equated the rate of adsorption to the rate of desorption of gas atoms or molecules on a smooth surface, where there is no interaction between adsorbed particles. Also, when a particle strikes a bare surface it adsorbs. When it strikes an occupied site on the surface it is reflected back into the gas phase.

The rate of adsorption is $k_a[A](1-\theta)$ and the rate of desorption is $k_{-d}\,\theta$, where θ is the fraction of surface covered. Equating these and defining $K = k_a/k_{-d}$ as an equilibrium constant for the adsorption process, we derive:

$$\theta = \frac{K[A]}{1+K[A]} \tag{1.6}$$

Here $[A]$ is the gas phase concentration of the molecular species A. For low values of $[A]$, the surface coverage varies linearly with the concentration $[A]$ or the gas pressure. For higher pressures, the coverage approaches unity. This relationship is illustrated in Figure 1.2 below.

Here we see the adsorption behavior of O_2 on Rh(111) at 335°K for various exposures. The unit of exposure is in

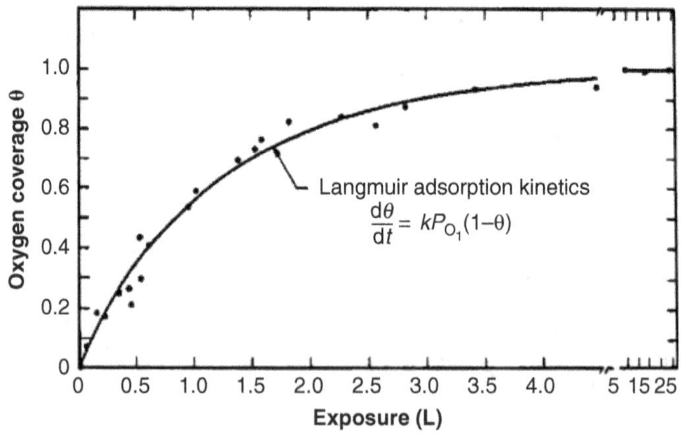

Figure 1.2 Adsorption kinetics of O_2 on Rh(111) [9].

Langmuirs (L = 10^{-6} Torr sec). As we can see, the experimental data appears to agree reasonably well with the prediction of the Langmuir isotherm. The surface coverage increases rapidly at initial exposure, as the incident flux increases with pressure, because there are many available sites for adsorption. As the coverage approaches unity, the rate of adsorption slows down, because there are fewer and fewer sites available for species to adsorb to.

Although the study of adsorption isotherms is an important area of investigation, especially for the study of surface catalysis, it plays a lesser role in ALD phenomena. That is, it is in general straightforward to dose a surface at a high enough pressure to form essentially a complete monolayer of the first reactant by adsorption. For example, nitrogen at one millitorr pressure, at room temperature, will have a number density of ~ 3×10^{13} molecules/cm^3, and an impingement rate onto a surface of ~ 4×10^{17} molecules/sec-cm^2. With a typical monolayer density of 10^{15} molecules/cm^2 the impingement rate is ~ 400 monolayers/sec. Therefore, there will be more than enough of a flux of gas phase species to the solid surface for a complete monolayer with 100% coverage to be formed.

1.2.2 Chemisorption

Chemisorption is similar to physical adsorption except that the reactant being adsorbed forms a chemical bond with atoms on the surface. With this type of bond the forces holding the adsorbed layer in place are on the order of chemical bonds between atoms in a molecule. The significance of this fact for thermal ALD is that the surface reaction of interest between the two reactants may require a moderately high temperature in order to obtain a rapid reaction. In this case, the first reactant will have to adsorb and remain bound to the surface until the reaction is complete. Physisorbed reactants will be desorbed by the higher temperatures, because of their weak bonding strength.

The initial work on chemisorbed monolayers on a solid surface was carried out by Langmuir [8]. He studied the adsorption of oxygen onto hot tungsten surfaces, and observed that a chemical bond was formed between the tungsten and the oxygen, and it was not tungsten oxide. Since a chemical reaction is involved, there are a large variety of chemisorption phenomena. For example, when H_2 chemisorbs to a surface it frequently will dissociate on the surface. Many other examples are reported in the literature [10].

Another important consequence of chemisorption for ALD, is that as a reactant bonds to the surface it will, in general, form just a single monolayer. Additional reactants arriving at the monolayer coated surface would have to bond to it by van der Waals forces, and the surface temperature would discourage this. Therefore, chemisorption bonding invariably leads to the desired single monolayer of reactant. It has also been pointed out [8] that chemisorption is often associated with an activation energy, so that unless the temperature is high enough the chemisorption bonding could proceed slowly. This is why chemisorption is often referred to as activated adsorption.

Finally, one has to ask the question as to how do we know which reactants to choose to be assured that a single monolayer will form on a surface. Clearly, there has been a large amount of research related to catalysis that deals with this subject, and there are many experimental and theoretical tools one can use to make predictions [10]. For ALD, however, we have a catalog of reactants that will chemisorb a single monolayer, as desired. Every paper that reports a successful ALD experiment identifies a reactant that will behave this way. In general, the halogen and organometallic compounds seem to chemisorb as needed. As we review the literature in the remainder of this book, we will identify many of these reactants.

1.3 Atomic Layer Deposition (ALD)

The first work on Atomic Layer Deposition, then known as Molecular Layering, was carried out in Russia by Aleskovski and co-workers in the 1960s [11]. The field of ALD was then further developed by Suntola and his Finnish collaborators [12] in the 1970s, as a new technique for the growth of polycrystalline dielectric thin films with unique characteristics. The essence of the process is that two distinct reactants are separately exposed to the growth surface in turn, and the reaction occurs between a first monolayer of reactant and a second reactant to form a solid thin film product. The name Atomic Layer Epitaxy (ALE) was chosen by Suntola. In fact only some of this work involved epitaxial films, primarily for semiconductors such as silicon and GaAs (see Chapters 2 and 3). Therefore, the term generally used today to describe this process is Atomic Layer Deposition (ALD).

The original work reported by Suntola [12] was presented in a process patent, and described the formation of a compound film by exposing the surface, in turn, to two elemental species. We will refer to this work as "direct" ALD. Next, in a second patent [13]

Suntola proposed the use of compounds containing the elements of interest, and exposing them to a heated surface, in turn, to form the desired solid thin film. Since now a chemical reaction was required to form the solid film, surface temperature became an issue, and the concept of a "temperature window" was developed. Finally we will review some of the characteristics both of the films that can be deposited and the deposition process itself.

In recent years, there has been an explosion of research in ALD. Several review papers have been recently published [14–19] and there have been many conferences on the subject, notably the AVS-sponsored annual ALD conferences [20].

1.3.1 Thermal ALD Processes

1.3.1.1 Direct ALD

In the direct ALD process [7], the surface upon which a film is to be grown is subjected to atomic species rather than compounds. In this case, the substrate temperature has to be maintained at a value high enough to prevent condensation of the first reactant, and a monolayer of it must remain after the vapor is evacuated from the chamber. Referring to our earlier discussion, this is a case of physisorption. The surface is then exposed to the second elemental species, which reacts with the first elemental monolayer to form a compound material. This process is then repeated as many times as necessary to grow a film of the desired thickness. In this early reference the authors describe the growth of ZnS, SnO_2, and GaP by this technique.

Also, several reactor configurations are described, and one of these is illustrated in Figure 1.3 below. This reactor configuration is typical of many that have been reported over the years. It shows a stationary substrate being dosed with two different elemental species, in turn, by opening and closing two valves. A vacuum pump evacuates the chamber after each pulse.

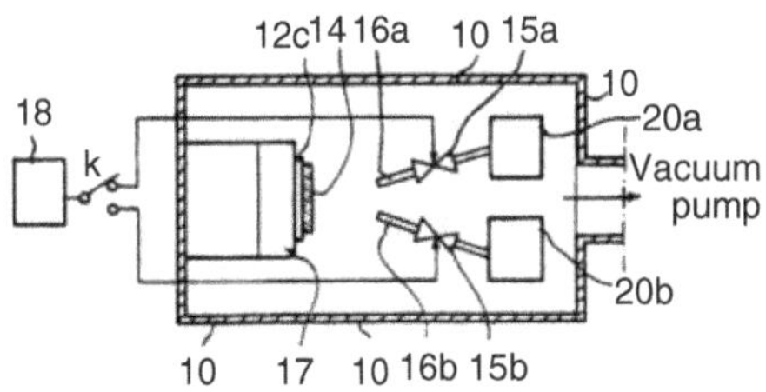

Figure 1.3 ALD reactor [12].

1.3.1.2 ALD with Compounds

Following his initial studies of ALD using the direct process, Suntola suggested that the same process could be carried out using chemical compounds rather than elemental species [8]. The difficulty in using elemental species to form compounds for ALD is that some of the elements desired in the final film have a low vapor pressure unless heated to high temperatures. Because of this, it was difficult to control the physisorption of a "self-limiting" monolayer of these atoms to the growing surface. If, on the other hand, a higher vapor pressure molecular species could be found, which easily chemisorbs to the surface, a self limiting monolayer is more likely to form and remain.

Accordingly, the concept of using halogen molecular species (e.g. $TiCl_4$, $AlCl_3$) or organometallic compounds (e.g. trimethylaluminum or TMA) was proposed. Since the original proposal by Suntola, many studies have been reported, and will be reviewed in the later portions of this book.

1.3.1.3 Temperature Window

For thermal ALD processes there are temperature constraints on the feasibility of successfully carrying out a particular deposition [21]. A schematic representation of this situation is shown in Figure 1.4.

On the vertical scale we show deposition rate versus temperature on the horizontal axis. If the temperature is very high, then the first chemical reactant may decompose on the surface before having time to react with the second reactant. In this case the growth rate would be higher than one would expect from an ALD process.

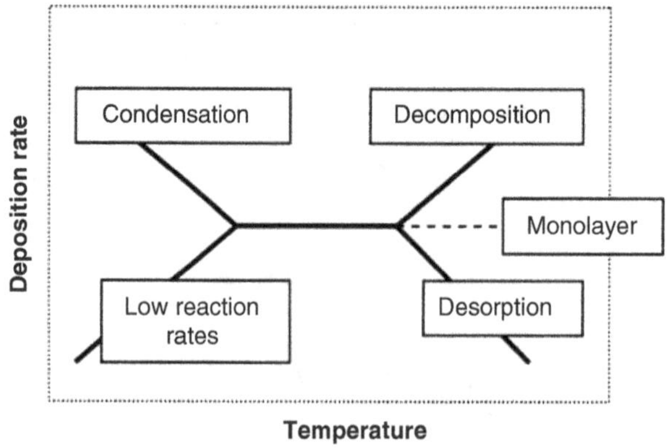

Figure 1.4 Temperature window for ALD.

Alternately, if the first precursor is stable it may still desorb from the surface before having a chance to react with the second reactant. Under this circumstance the growth rate would be less than expected. This would be more likely with a physisorbed first reactant than a chemisorbed one. On the other hand, if the temperature is too low, we may adsorb more than one monolayer per cycle (or even condense a liquid or solid on the surface), and the deposition rate would be higher than expected. Finally, if the temperature is too low, the reaction rate may be so slow that the reaction time may be too long compared to a practical cycle time (e.g. hours rather than seconds). In this case, there may not be enough time for a complete monolayer to be reacted.

Although many chemical systems are amenable to thermal ALD, as will be seen in the remainder of this book, there are some for which the temperature window is closed. For example, if an attempt were made to deposit titanium metal from $TiCl_4$ and H_2 it would be recognized that the temperature required to activate this reaction would be about 2000°C, and at this temperature any $TiCl_4$ reactant that would have adsorbed would have promptly desorbed, preventing an ALD type of growth. One way to remove the restriction of the temperature window would be to use a highly reactive radical as one of the reactants. This will be the subject of Section 1.3.2.

As noted earlier, when depositing films by CVD the most reactive set of reactants may be desirable in terms of lowering the reaction temperature, but they also cause many other difficulties because they are mixed in the gas phase and can react there. In ALD we again seek the most reactive set of reactants, however now there is little disadvantage in their use. This is because the only time they interact is when they are on the surface of the substrate, and there we want the fastest reaction possible at the lowest temperature practical. One example of this phenomenon can be seen from the ALD deposition of aluminum oxide, which will be covered in more detail in Chapter 4. When Al_2O_3 is grown from TMA and water vapor we get good quality films at a substrate temperature of 200°C. If we want to grow at lower temperature, film growth still takes place although with some increase of impurities such as C but the limiting factor becomes the rate at which excess water vapour can desorb from the surface. Purge times for water of less than 1 s at higher temperatures can increase to 180 s at a temperature as low as 33 °C [22]. If, however, a more reactive oxidiser is used, such as ozone or oxygen radicals produced in a plasma assisted process then short purge times can be maintained down to room temperature. So, in general, when choosing reactants for an ALD process one would select the most reactive pair available provided that each reactant has sufficient thermal stability not to spontaneously decompose in the gas phase.

1.3.1.4 Deposition Rates

As will be seen as we review the various films deposited by ALD by many authors, the rate of film deposition varies widely. In most cases the rate represents a deposition rate that is less than a monolayer of the product film per cycle. For example, we may adsorb a self-limiting single monolayer of the first reactant. However, since these are often large molecules with perhaps only one element that we wish to leave in the deposited film, there are not enough atoms of that element to form a complete single monolayer of the product film. This is often referred to as the "steric hindrance" effect. For example, in the case of deposition of Al_2O_3 using TMA we show (see Chapter 4) that the first monolayer of the large ~ 6 Å diameter TMA molecule only contains 20% of the aluminum atoms required to form a complete monolayer of Al_2O_3. Thus only approximately one fifth of a monolayer of Al_2O_3 can be grown per process cycle. For a 3.5 Å diameter aluminum atom this yields a growth rate of 0.7 Å per cycle, which is close to the values reported in the literature.

From the above approximate calculation, we can see deposition rates for metal oxide materials will be determined largely by the nature of the metal-containing reactant. In cases where it is a large molecule, we can anticipate that we will not obtain enough elemental species to grow a full monolayer of product film during each cycle, and this will give slower deposition rates.

1.3.1.5 Characteristics of ALD Processes

The ALD process has certain inherent advantages, in terms of thin film deposition, compared to all other available thin film deposition techniques. First of all, we should note that, in general, ALD film deposition does not depend on the several items of critical importance in CVD deposition. For example, neither temperature nor flow uniformity are of great importance in determining the quality of the films formed. As long as the temperature lies within the "temperature window", the uniformity of the local temperature on the substrate surface should not influence the film characteristics. Similarly, the local concentration of reactant molecules does not need to be uniform provided that all parts of the surface receive at least enough molecular flux to form a monolayer of the reactant.. For the above reasons, it is not difficult to deposit very uniform films (in thickness and in chemistry) over large area substrates by the ALD process. At the same time, because the films are deposited a layer at a time, they will tend to be very conformal over uneven topography.

ALD film growth rates are in general significantly lower than for films deposited by PVD, that is evaporation or sputtering. However, where careful control of film properties, such as morphology and conformality, are desired ALD may be the method of choice. Specifically in relation to semiconductor wafer coating, the currently preferred sputter deposition schemes have reached the limits of their ability to provide a satisfactory via liner coating (Ti, TiN) for advanced device generations with submicron size high aspect ratio vias. In this application, conformality, film thickness uniformity and the need for less than 100 Å thick films, are all serious issues for PVD. It is already the case that ALD is the method of choice for achieving conformal deposition of gate dielectric layers in microelectronics and for seed layers for metallisation.

As a final point, it is possible, by ALD, to grow multilayer thin films known as nanolaminates. That is we can tailor the film characteristics by depositing alternate layers of different materials, sequentially. In other words, one can deposit "designer" films as required.

A non-exhaustive list of some of the thin film materials that have been deposited by ALD is shown in Table 1.1. As can be seen, this

Table 1.1 A non-exhaustive list of typical thin films deposited by ALD.

Oxides	Nitrides	Sulfides	Semiconductors	Elements
Al_2O_3	TiN	SrS	GaAs	Si
In_2O_3	NbN	CaS	Si	Ge
ZrO_2	TaN	BaS	InAs	Cu
HfO_2	Ta_3N_5		InP	Mo
Ta_2O_5	MoN		GaP	W
SiO_2	WN		InGaP	Ta
TiO_2	BN			Ti
SnO_2				Pt
Nb_2O_5				Ru
ZnO				Ni

process can deposit a very wide variety of materials, and in fact, many more processes are continuously being developed. All of these films have been deposited from a large variety of compounds. By and large, the first reactant is either a halogen or an organometallic. Recently there have been many new metallic precursors developed of increasing molecular complexity; a short selection of typical precursors is shown in Table 1.2. When an oxide is being deposited, the second reactant is often water vapor or ozone. Nitrides have been formed using ammonia as the second reactant although other more reactive nitrogen bearing species have been used. For sulfides, hydrogen sulfide or liquid organo-sulphur compunds are generally chosen as the second reactant. Finally, for compound semiconductors, arsine and phosphorous vapor are frequently selected although liquid organo-phopsphorus or organo-arsenic compounds have been used for their ease of handling. A growing area of study is into metal-organic films such as the "alucones" where an aluminum precursor reacts with an organic molecule such as glycol to produce an alternating metal and organic structure [23].

1.3.2 Radical Enhanced ALD (REALD)

As outlined in section 1.3.1 and illustrated in Figure 1.4, the ALD process takes place within a "temperature window" which depends on the stability of the precursor chemicals and their reaction with the surface. As shown in Figure 1.5, if the temperature at which the reaction starts to take place is too high then the temperature window gets shorter and may, in fact, disappear. Therefore, for many thermal processes, there is no acceptable temperature window, and no suitable deposition can be achieved. The approach where one of the reactants is an energetic radical can maximize this "temperature window" and even, enable processes at room temperature in some cases. This is what we will refer to as Radical enhanced ALD (REALD).

A radical is an atom or molecular fragment that is unstable and releases a great deal of energy when it returns to a stable configuration. For example, when a hydrogen molecule (H_2) is decomposed into two atoms of hydrogen, they release the energy of recombination when they recombine. When such recombination occurs on a surface, the energy released can participate in a chemical reaction with another molecule that has adsorbed to the surface. However they are created, free radicals when used in an ALD process, can provide a versatile route for process and thin film application

Table 1.2 Typical precursor compounds used as metal sources in ALD processes.

Alkyls	TMA trimethyl aluminum	$Al(CH_3)_3$
	TEA triethyl aluminum	$Al(C_2H_5)_3$
	DEZ diethyl zinc	$Zn(C_2H_5)_2$
Alkoxides	Titanium ethoxide	$Ta(OC_2H_5)_5$
	Titanium isopropoxide	$Ti(OCH(CH_3)_3)_4$
Alkyl amides	TDMAT tetrakis(dimethylamino)titanium	$Ti[N(CH_3)_2]_4$
	TEMAZ Tetrakis(ethylmethylamino)zirconium	$Zr(N(CH_3)(CH_2CH_3))_4$
Amidinates	Bis(N,N′-di-i-propylacetamidinato) cobalt(II)	$(C_8H_{17}N_2)_2Co$;
Diketonates	Cu(hfac) Copper(II) hexafluoroacetylacetonate,	$Cu(C_5HF_6O_2)_2$
	Ni(acac) Nickel(II) acetylacetonate	$Ni(C_5H_7O_2)_2$
Carbonyls	Iron pentacarbonyl,	$Fe(CO)_5$
	Chromium hexacarbonyl	$Cr(CO)_6$
Cyclopentadienyls	Bis(ethylcyclopentadienyl)ruthenium(II)	$C_7H_9RuC_7H_9$
	Bis(ethylcyclopentadienyl)magnesium	$Mg(C_5H_4C_2H_5)_2$
Halogens	Titanium chloride	$TiCl_4$
	Titanium iodide	TiI_4
	Zirconium chloride	$ZrCl_4$
	Tantalum chloride	$TaCl_5$
	Hafnium chloride	$HfCl_4$
	Tin chloride	$SnCl_4$

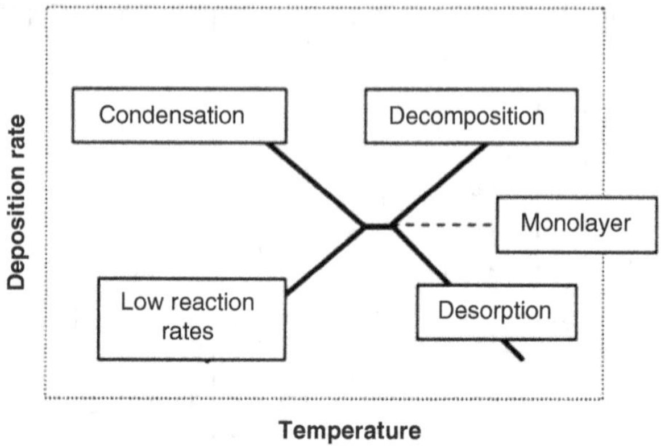

Figure 1.5 Deposition temperature window for ALD for reactions with slow reaction rates.

development. The benefits that the radical enhanced ALD process can offer are reduced deposition temperatures, increased fragmentation of the precursor molecules, wider selection of precursors for the thin film synthesis, and improved material properties. In addition, if a plasma is used to create the free radicals they may be electrically charged and provide further enhancement of the process through ion bombardment of substrates.

Very early in the development of ALD processes, various techniques were explored for the activation of surface reactions by non thermal means. Beginning in the mid 1980's a major effort was started to deposit thin epitaxial silicon (see Chapter 2), and a number of III-V semiconductor films (GaAs and others, see Chapter 3). These efforts were supported with substantial US Government research funding. The motivation was to try to construct super lattices of different single crystal semiconductor materials. By the mid 1990's it became clear that this was a very difficult challenge, and funding was substantially reduced. None the less a number of novel approaches were explored during these attempts. Since at that time commercial development of semiconductor devices (microprocessors and memory chips) was only requiring fairly thick material layers, the prevailing wisdom was that ALD was simply too slow to be a commercially viable process. As a result, this area of research languished. Of course, research continued on thermal ALD of oxides, nitrides, etc., but not for the semiconductor industry.

In Chapters 2 and 3 we mention the attempts to enhance the growth of Si and GaAs using optical stimulation, but report that these efforts were not particularly successful. Of greater interest was the use of a hot tungsten filament to generate hydrogen atoms (radicals) from hydrogen gas, and the partial success in growing epitaxial silicon and germanium in this way (see Chapter 2). Finally, in 1990 an attempt to grow GaAs was reported using a glow discharge to create hydrogen atomic radicals by dissociating hydrogen gas. These radicals were used to stimulate the reaction between TMGa and arsine (see Chapter 3) and grow GaAs. Unfortunately, this paper was the only one published on this subject, and did not receive any attention from the semiconductor industry. In 1999 the REALD concept was extended to all materials, other than semiconductors, by Sherman [24]. A subsequent study at IBM research labs [25] in 2000 then demonstrated, for the first time, that a REALD process using hydrogen plasma could deposit pure Ti and Ta (see Chapter 6). Coinciding with this development, the semiconductor industry finally became interested in any film deposition method that could deposit very thin films with the characteristics that ALD offered. Once the very large commercial semiconductor industry became committed to this new approach, R & D efforts began to ramp up. As can be seen in Figure 1.6, the research effort has been growing rapidly, especially for REALD processes, in recent years.

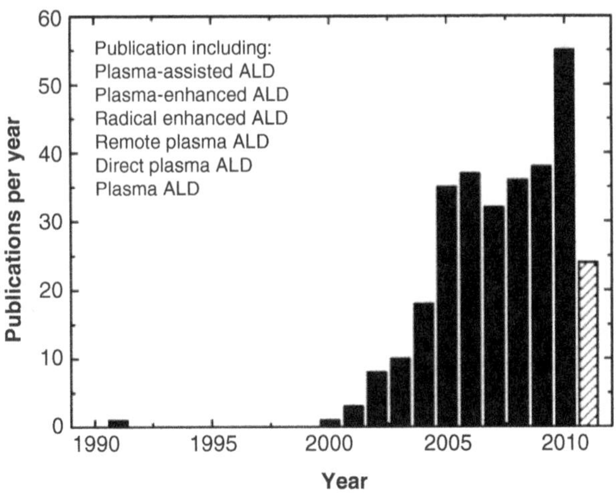

Figure 1.6 Number of publications per year on the subject of plasma-assisted ALD [26]. *Reprinted with permission from H. B. Profijt, S. E. Potts, M. C. M van de Sanden, and W. M. M. Kessels, J. Vac. Sci. Technol. A 29, 050801 © 2011, American Vacuum Society.*

Of all of the ways that energetic radicals could be created, the simplest is to generate them in a glow discharge created in a low pressure reactor chamber. Therefore, all commercial REALD reactors are really PEALD systems. These systems are distinguished by the various plasma sources available such as capacitive discharges, discharges created by inductive solenoid coils, microwave discharges, etc. Further if ion bombardment is required the reactor can be configured to add such a feature. Similarly some systems arrange to have the primary discharge spaced some distance from the part being coated. Some of the different reactor configurations are presented schematically in Figure 1.7 [26]. Interestingly all of these resemble PECVD reactors that have been studied in research labs and used commercially for many years.

In fact several of the reactors described above are illustrated in the present text. For example, in Chapter 2 (Figure 2.4), we have

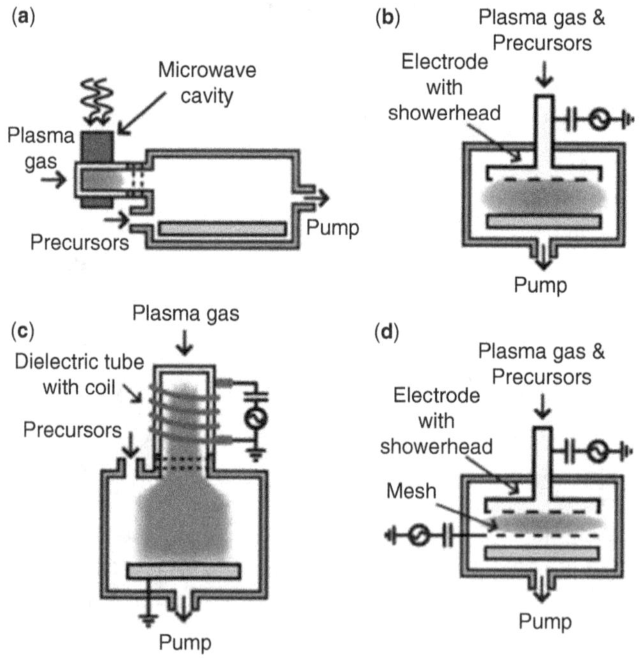

Figure 1.7 Schematic illustration of plasma-assisted ALD reactor configurations: a) radical- enhanced ALD, b) direct plasma-assisted ALD, c) remote plasma ALD and d) direct plasma reactor with mesh [26]. *Reprinted with permission from H. B. Profijt, S. E. Potts, M. C. M van de Sanden, and W. M. M. Kessels, J. Vac. Sci. Technol. A 29, 050801 © 2011, American Vacuum Society.*

a reactor used to deposit elemental silicon using hydrogen atoms created by catalytic decomposition of hydrogen gas on a hot tungsten filament. In the same Chapter (Figure 2.8) we see a reactor that uses a remote inductively excited discharge to produce a Helium plasma.

1.3.2.1 *Definition of REALD and PEALD*

The actual configuration and placement of the plasma source in the ALD reactor determines the performance of the source in respect to the radical generation efficiency and what other effects plasma may have on ALD film growth during the plasma step. In the radical enhanced ALD configuration, where the plasma source is placed relatively far from the ALD reaction chamber (Figure 1.6a), the flux of plasma species to the substrate surface will be determined e.g. by the tubing length and material affecting the surface collisions and recombinations of plasma species. This type of configuration is prone to significantly reduced radical flux reaching the substrate surface and will consequently require longer radical exposure times. On the other hand the substrate surface will not be exposed, at least not to a large extent, to plasma photons, ions and electrons.

In the direct plasma configuration, Figure 1.6b, the plasma is generated by capacitive coupling. A common plasma source in this type of configuration is radio frequency power (typically 13.56 MHz). In the direct plasma configuration one electrode is powered and the other one, where the substrate is also positioned, is grounded. Since the distance between the plasma source and the substrate in this type of configuration is small, it can result in very high fluxes of radicals and other plasma species towards the substrate surface. Consequently it allows the reduction of plasma exposure time and the design is amenable also to large area deposition.

In the remote plasma configuration, Figure 1.6c, the plasma is located in the same chamber as with the substrate but still remotely from it, such that the substrate is not involved in the plasma generation as in the case of the direct plasma configuration. Inductively coupled plasma is typically used in this configuration. The benefits of the remote plasma configuration are the higher flux of the radicals (compared to radical enhanced configuration) towards the substrate surface. The design allows the control of plasma conditions independently from the substrate conditions and gives also more control for adjusting the plasma properties compared to direct plasma configuration [26]. Uncontrolled plasma conditions can lead to plasma damage of the substrate material or grown film. This has been demonstrated e.g. with photo sensitive polymer (PMMA), where uncontrolled plasma conditions using direct plasma configuration led to polymer degradation through intensive VUV radiation [32], and with Al_2O_3 passivation layers on Si, where uncontrolled plasma conditions by using remote plasma configuration created electrical defects in the films also due to the VUV photons [33]. However in all plasma configurations the plasma induced effects depend on plasma source geometry, source type, plasma power, the gas composition, the gas pressure and flow conditions [26, 34].

1.3.3 Spatial ALD (SALD)

In conventional, or batch ALD, the process sequence occurs by delivering the precursors to the stationary substrate in a time separated series of pulses. In order to remove excess precursor and by-product materials from the reaction chamber, these precursor pulses must be separated by pulses of purge gas thus giving the familiar precursor, purge, precursor, purge gas sequence. This leads to certain process characteristics:

i. Accurate control of the gas pulsing process is required.
ii. The gas pulse lengths have to be long enough to ensure that enough material fills the reaction chamber to form a monolayer on all surfaces.
iii. The purge and exhaust pulses have to be long enough to ensure that all remaining gas-phase precursors or reaction by-products are removed from the chamber.
iv. All the internal surfaces of the reaction chamber are coated as well as the substrate, since they are all exposed to the same process.

The chamber filling and emptying times require the pulse cycle to be longer than would be necessary if the process were controlled by the surface reactions. The deposition on the chamber walls means that there is the possibility of flaking and particle contamination of the substrate. This unwanted deposition also implies low utilisation efficiency for the precursors which can be very expensive for some processes.

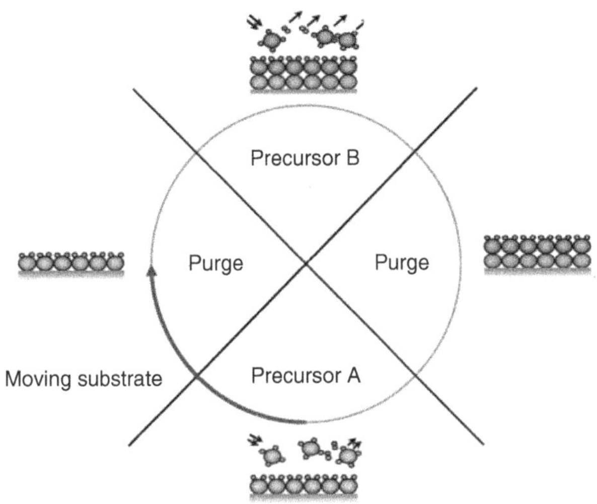

Figure 1.8 Schematic of spatial ALD cycle.

The same sequence of exposure of the substrate to the precursors and purge gases can, however, also be achieved by having constant flows of gases in spatially separate zones and physically moving the substrate between these zones as shown schematically in Figure 1.8. This process is known as spatial ALD and it is currently attracting much attention. Significant developments in process equipment are taking place since this new process allows the possibility of a continuous in-line process or roll-to-roll web coating which is necessary for many important market areas such as moisture diffusion barriers for flexible OLEDs and thin film photovoltaics. Other advantages of the spatial ALD process are that:

i. There is no time spent in filling and emptying the chamber between pulses since the gas flows are static in each zone. consequently, the speed of the deposition cycle is limited by the time it takes to move the substrate between gas zones and the speed of the surface reaction between the precursor and substrate.
ii. The physical separation of the precursors in different gas zones and the absence of any intermixing mean that only part of the system which is exposed to the sequence of precursors is the moving substrate. Therefore, the deposition process only occurs on the substrate which means that the efficiency of precursor utilisation can be

much higher and there is less likelihood of particulate contamination from the chamber walls due to flaking.

Although there has recently been increasing interest in spatial ALD, the concept goes back to an early patent by Suntola *et al.* in 1983 [35]. As shown in Figure 1.9, their design incorporated separate precursor zones, intermediate purge gas zones to act as gas separation regions and exhaust zones to ensure that there was no intermixing of gases. Other notable SALD designs which have been patented include a system with a linear multichannel gas zone head [36] with close spacing between the substrate and coating head to ensure adequate gas separation (Figure 1.10), and a web-coating system where the flexible substrate was led in serpentine manner

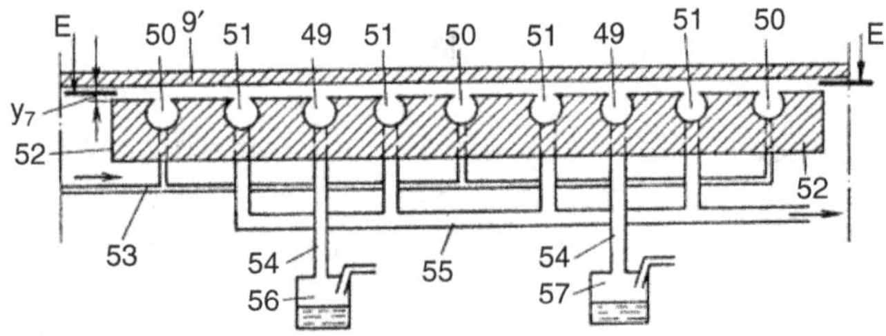

Figure 1.9 Initial proposed design for a spatial ALD system [35].

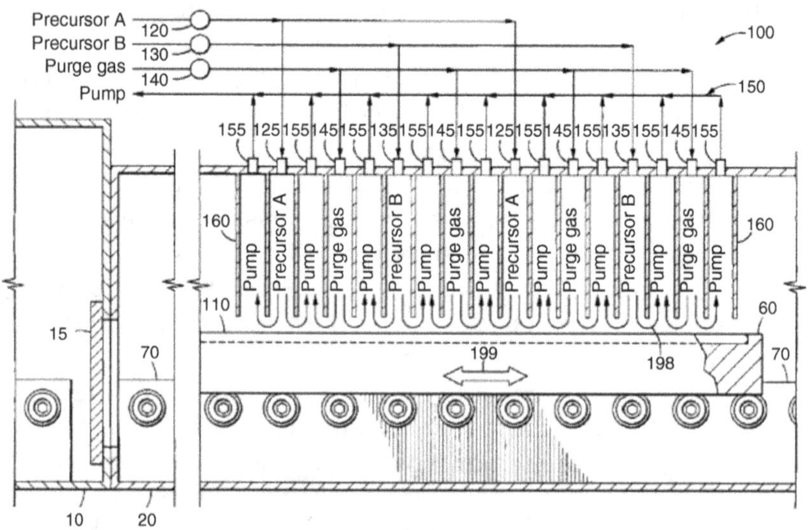

Figure 1.10 Design for a spatial ALD system [36].

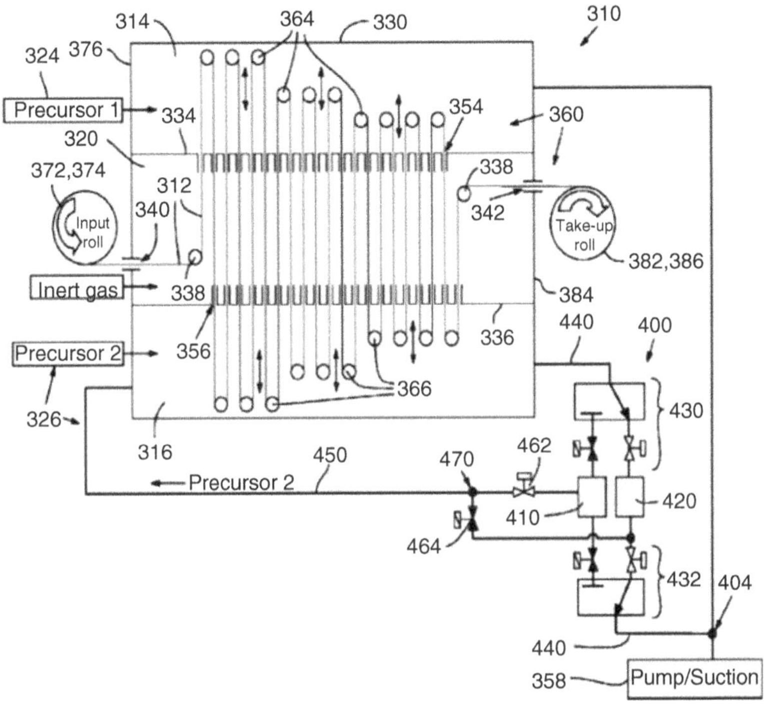

Figure 1.11 SALD design for flexible web coating [37].

through two separate precursor zones separated by a purge gas zone [37] (Figure 1.11). These two basic models types typify the systems which are currently under development. In addition, the various current reactor designs enable operation either at atmospheric pressure requiring no vacuum pumping or at a reduced pressure similar to conventional ALD. The current state of the art in SALD has been recently reviewed [38].

1.3.3.1 *Atmospheric Pressure SALD*

Atmospheric pressure spatial ALD systems have been based on the basic design of Yudovsky [36] (Figure 1.10). The various precursor, purge and exhaust ports are built into a coating head which moves relative to the substrate. There have been a number of such systems [39–41]. The critical factor in these is the need for very small separation between the coating head and the substrate (<100 μm) to ensure no gas intermixing. The pressure of the purge gas which isolates the precursor zone can itself be used to ensure the levitation

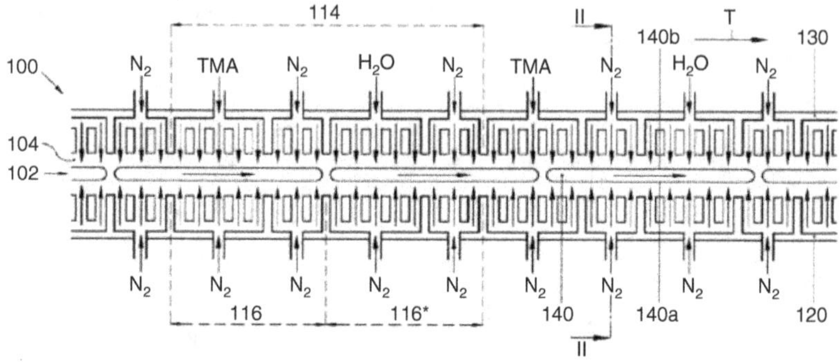

Figure 1.12 Schematic of linear track SALD system [44].

of the substrate above the head and to maintain its correct spacing. This gas levitation ensures very low friction movement between the substrate and the reactor and can control the substrate to head separation to be in the range of tens of micrometres. Results using this type of system were first reported by Levy et al. who subsequently demonstrated the deposition of aluminum oxide insulators and zinc oxide semiconductors [42] for thin film transistors by this method. A rotary system based on the gas bearing technology was developed by TNO in the Netherlands [40] which demonstrated high rates of deposition of aluminum oxide and a prototype linear in-line system based on this experimental work has been developed by Solaytec where the substrate can be reciprocated back and forth across a number of coating heads, suspended on a gas bearing [43].

An SALD tool for aluminum oxide backside passivation of silicon solar cells has been developed by Levitech [41]. In this system, silicon wafers are transported along a linear track by gas pressure levitation past multiple coating heads as shown in Figure 1.12 [44]. Good passivation performance was demonstrated with Al_2O_3 where 5–10 nm was achieved in ~1 s with a throughput of 3600 156 × 156 mm^2 wafers/hr [38].

1.3.3.2 *Reduced Pressure SALD*

Reduced pressure SALD systems have been developed for deposition on to flexible substrates with the aim of roll-to-roll coating. Two types of design have been demonstrated. Dickey et al. from Lotus Technology developed a system based on their patent [37] where they have deposited on 100 mm width polymer webs using the TMA/water Al_2O_3 process and $TiCl_4$/water TiO_2 process. These have demonstrated good diffusion barrier properties [45]. They have also used a plasma oxidation process for the oxidising step in order to remove water as the precursor [38]. This system is shown in Figure 1.13.

An alternative approach using the multi-head design as in the atmospheric pressure systems but configuring it to enable coating of flexible materials has been reported by Maydannik et al. [46, 47]. In this system, a cylindrical reaction chamber has been built with the various gas and exhaust zones round the circumference. The flexible substrate is mounted on a rotating drum such that one rotation is

Figure 1.13 a) Schematic representation and b) image of the Lotus Applied Technology SALD reactor [38]. *Reprinted with permission from P. Poodt, D. C. Cameron, E. Dickey, S. M. George, V. Kuznetsov, G. N. Parsons, F. Roozeboom, G. Sundaram, A. Vermeer, J. Vac. Sci. Technol. A 30, 010802 © 2012, American Vacuum Society*

equivalent to one ALD cycle. This is shown in Figure 1.14. A photograph of this reactor configuration in Beneq's TFS 200R ALD tool is shown in Figure 1.15. Ultra-close substrate to coated head separation is not required in this case because of the lower pressure and the design of the rotating drum which helps to confine the reactant gases. A commercial roll-to-roll system, Beneq WCS 500, capable of coating

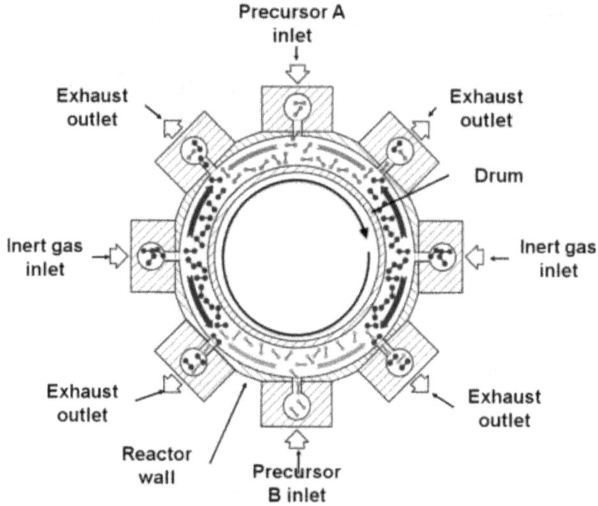

Figure 1.14 SALD chamber for flexible substrates [46]. *Reprinted from Chemical Engineering Journal, 171, P. S. Maydannik, T. O. Kääriäinen, D. C. Cameron, An atomic layer deposition process for moving flexible substrates, 345, © 2011, with permission from Elsevier.*

Figure 1.15 A photograph of SALD chamber for flexible substrates [38]. *Reprinted with permission from P. Poodt, D. C. Cameron, E. Dickey, S. M. George, V. Kuznetsov, G. N. Parsons, F. Roozeboom, G. Sundaram, A. Vermeer, J. Vac. Sci. Technol. A 30, 010802 © 2012, American Vacuum Society.*

Figure 1.16 A photograph of Beneq's WCS 500 roll-to-roll ALD reactor. *Photo courtesy and Copyright Beneq Oy 2013.*

500 mm wide web based on this principle has been developed by Beneq Oy. The overview of vacuum vessel with feed-in and take-up reels, and processing drum in the middle is shown in Figure 1.16. ALD coating head is situated below the processing drum.

References

1. A. Sherman, *Chemical Vapor Deposition*, Noyes Publications., New Jersey, 1987.
2. J. Bloem and L.J. Giling, Mechanisms of the Chemical Vapor Deposition of Silicon, in *Current Topics in Materials Science*, Vol. I, ed. Kaldis, E., North-Holland Publishing, 1978.
3. J Garcia, R Pitonak, R Weissenbacher and A Kopf, *Surface & Coatings Technology* 205, 2322 (2010).
4. G L Puma, A Bono, D Krishnaiah and J C Collin, *J Hazardous Materials* 157, 209 (2008).
5. M Kumar and Y Ando, *J Nanoscience Nanotechnology* 10, 3739 (2010).
6. W C Ren,, L B Gao, L PMa, and H M Cheng, *New Carbon Materials* 26,71 (2011).
7. M E Alf, A Asatekin, M C Barr, S H Baxamusa, H Chelawat, G Ozaydin-Ince, C D Petruczok, R Sreenivasan, W E Tenhaeff, N J Trujillo, S Vaddiraju, J J Xu and K K Gleason, *Advanced Materials* 22, 1993 (2010).
8. S. Glasstone, *Textbook of Physical Chemistry*, 2nd ed., D. Van Nostrand, Princeton, NJ, 1946.

9. A. Zangwill, *Physics at Surfaces*, Cambridge University Press, Cambridge, 1988.
10. K.J. Laidler, *Chemical Kinetics*, 3rd ed., Harper Collins, NY,1987.
11. S. I. Kolzov,V. B. Aleskovski, *Zh Prikl Khim*. 42, 1210 (1968).
12. T. Suntola and J. Antson, *U.S. Patent No*. 4,058,430 (1977).
13. T.S. Suntola, A.J. Pakkala and S.G. Lindfors, *U.S. Patent No*.4,389,973 (1983).
14. H. B. Profijt, S. E. Potts, M. C. M. van de Sanden and W. M. M. Kessels, J. Vac. Sc. *Technol. A* 29, 0508110 (2011).
15. S. M. George, *Chem. Rev*. 110, 111 (2010).
16. M. Knez, N. Kornelius and L.Niinistö, *Adv. Mater*. 19, 3425 (2007).
17. R. L. Puurunen, J. Appl. Phys. 97, 121301 (2005).
18. M. Leskelä and M. Ritala, *Angewandte Chemie-Intl*. Ed. 42, 5548 (2003).
19. H.Kim and P. C. McIntyre, *J. Korean Phys. Soc*. 48, 5 (2006).
20. www.ald-avs.org
21. M. Leskelä, Acta Polytechn. *Scand., Ser. Chem. Techn*., 195, 67 (1990).
22. M. D. Groner, F. H. Fabreguette, J. W. Elam and S. M. George, *Chem. Mater*. 16, 639 (2004).
23. A. A, Dameron, D. Seghete, B. B. Burton, S, D, Davidson, A. S. Cavanagh, J. A. Bertrand, S. M. George, *Chem. Mater*. 20, 3315 (2008).
24. A. Sherman, *US Patent* 5,916,365 June 29, 1999
25. S. M. Rossnagel, A. Sherman, F. J. Turner, *Vac. Sci. Technol. B* 18(4) 2000.
26. H. B. Profijt, S. E. Potts, M. C. M van de Sanden, and W. M. M. Kessels, *J. Vac. Sci. Technol. A* 29(5), 050801 (2011).
27. S. E. Potts, W. Keuning, E. Langereis, G. Dingemans, M. C. M. van de Sanden, and W. M. M. Kessels, *J. Electrochem. Soc*. 157 (2010).
28. A. Niskanen, T. Hatanpää, K. Arstila, M. Leskelä and M. Ritala, Chem. Vap. *Deposition* 13, 408 (2007).
29. C-W. Jeong, B-I. Lee, S-K. Joo, *Mater. Sci. Eng*. C 16, 59 (2001).
30. Y. J. Lee, S-W.Kang, *J. Vac. Sci. Technol. A* 20(6) 1983–1988 (2002).
31. E. R. Dickey and W. A. Barrow, 53rd Annual Tech. Conf. Proc. Soc. Vacuum Coaters p.558 (2010).
32. T. O. Kääriäinen, S. Lehti, M.- L. Kääriäinen, and D. C. Cameron, *Surf. Coatings Technol*. 205, S475 (2011).
33. H. B. Profijt, P. Kudlacek, M. C. M. van de Sanden, and W. M. M. Kessels, *J. Electron. Soc*. 158 (2011).
34. S. B. S. Heil, J. L. van Hemmen, C. J. Hodson, N. Singh, J. H. Klootwijk, F. Roozeboom, M. C. M. van de Sanden, and W. M. M. Kessels, *J. Vac. Sci. Technol. A*, 25 (2007).
35. T. S. Suntola, A. J. Pakkala and S. G. Lindfors, *US patent No*. 4,413,022 (1983).
36. J. Yudovsky, *U.S. Patent No*. US2004067641A1 (2004).
37. E. R. Dickey and W. A. Barrow, *U.S. Patent No*. US2007224348A1 (2007).

38. P. Poodt, D. C. Cameron, E. Dickey, S. M. George, V. Kuznetsov, G. N. Parsons, F. Roozeboom, G. Sundaram and A. Vermeer, *J. Vac. Sci. Technol. A* 30, 010802 (2012).
39. D. H. Levy, D. C. Freeman, S. F. Nelson, P. J. Cowdery-Corvan and L. M. Irving, *Appl. Phys. Letts.* 92, 192101 (2008).
40. P. Poodt, A. Lankhorst, F. Roozeboom, V. Tiba, K. Spee, D. Maas and A. Vermeer, *Adv. Mater.* 22, 3564 (2010).
41. V. Kuznetsov, *11th Int. Conf. on Atomic Later Deposition*, Cambridge, (2011).
42. D. H. Levy, S. F. Nelson and D. Freeman, *J. Display Technol.* 5, 484 (2009).
43. B. Vermang, A. Rothschild, A. racz, J. John, J. Poortmans, R. mertens, P. Poodt, V. Tiba and F. Roozeboom, *Prog. Photovoltaics* 19,733 (2011).
44. E. H. A. Granneman and V. Kuznetsov, *Netherlands Patent No.* 2005049.
45. E. R. Dickey and W. A. Barrow, *52nd Annual Tech. Conf. Proc.* Soc. Vacuum Coaters , p. 720 (2009).
46. P. S. Maydannik, T. O. Kääriäinen and D. C. Cameron, *Chem. Eng. J.* 171, 345 (2011).
47. P. S. Maydannik, T. O. Kääriäinen and D. C. Cameron, *J. Vac. Sc. Technol. A* 30, 01A122 (2012).

2

Elemental Semiconductor Epitaxial Films

2.1 Epitaxial Silicon

Application of the ALD concept to the deposition of epitaxial thin films of elemental semiconductors (e.g. silicon and germanium) could have a potential impact on the commercial manufacture of integrated circuits, and has therefore been actively researched for some years. However, it is significantly more difficult to achieve an elemental epitaxial film, than it is to deposit the amorphous and polycrystalline compound films that were originally developed. In addition, investigations of the deposition by ALD of epitaxial III-V compound semiconductor films have been reported, and these developments will be reviewed in the next chapter. The number of papers on ALD of semiconductor films is, however, small when compared to those on dielectrics and other non-semiconductor materials reflecting the difficulties involved.

The reason it is so difficult to deposit an epitaxial film, by ALD, is that we have to replicate the single crystal structure that we are depositing on to. When we deposit, for example, silicon on silicon, this is referred to as homoepitaxy. If we deposit silicon onto a germanium single crystal substrate, this would be referred to as heteroepitaxy. It would be ideal if the precursor monolayer had enough silicon atoms available to allow a single layer of single crystal material (e. g. 6.8×10^{14} atoms/cm^2 for Si(001)) to be formed on the surface. Since this will not, in general, be the case, the substrate would have to be hot enough to allow some atom diffusion to single crystal sites.

The first step in depositing silicon or germanium by ALD is to choose a first reactant that contains one of these elements, and that forms a self-limiting monolayer on a substrate. To date, a number of suitable compounds have been used. For example, for silicon these include various silanes, halides, and even organic compounds. Once the first reactant has adsorbed onto the substrate, the problem becomes how to remove all species other than silicon. Just as important, we want to leave behind a complete monolayer of silicon per cycle. If it were possible to adsorb a SiH radical, for example, then there would be a good chance that we would have enough silicon atoms available to form a complete monolayer at each cycle. However, this is not a stable molecular species, so one ends up using larger molecules, and the one monolayer per cycle is harder to achieve due to steric hindrance.

The compound most widely studied for silicon ALD is dichlorosilane, and research with this material will be reviewed in the next section. Next, we will look at work done with other silicon compounds, and then cover some of the work done on germanium.

2.1.1 Dichlorosilane Processes

Among the precursors that could be used to deposit silicon by ALD are molecules with either chlorine or hydrogen, or a mixture of these. For example, one could consider $SiCl_4$, $SiHCl_3$, SiH_2Cl_2 and SiH_4. Higher silanes could also be tried (e.g. Si_2H_6 and Si_3H_8). The silanes will be covered in the next section. Most of the research reported has been with dichlorosilane, SiH_2Cl_2, and this will be the subject of the present section.

2.1.1.1 Using Molecular Hydrogen

The fundamental problem with trying to grow epitaxial silicon is the fact that a bare single crystal silicon surface has dangling bonds exposed. If after the first precursor layer has been reacted, these dangling bonds are terminated by either a chlorine or hydrogen atom, the surface will be passivated, and it will not be possible to chemisorb the next monolayer of precursor molecules. When a monolayer of molecules containing hydrogen or chlorine is reacted by a second reactant, there is indeed a high probability that some chlorine and/or hydrogen will be released and passivate the surface. So the trick is to find a way to remove any hydrogen and chlorine that might be remaining on the surface, after each reactant cycle.

The first reported studies of silicon deposition by ALD were from experiments with SiH_2Cl_2 and H_2 at high temperatures [1, 2]. The reaction scheme, which is highly simplified, is shown below:

$$SiH_2Cl_2 \rightarrow SiCl_2 + H_2 \tag{2.1}$$

$$SiCl_2 + H_2 \rightarrow Si(s) + 2HCl \tag{2.2}$$

Deposition was carried out at temperatures ranging from 775°–890°C, with a base pressure of less than 10^{-9} torr. The reactor used is shown schematically in Figure 2.1. The substrate is heated by an infrared lamp, and its temperature is monitored with a pyrometer.

The timing sequence is shown in Figure 2.2. Each gas is admitted to the closed volume of the reactor, allowed to remain for a period of time, and then evacuated before the next gas pulse is admitted. For the example of Figure 2.2, SiH_2Cl_2 is placed in the chamber at a

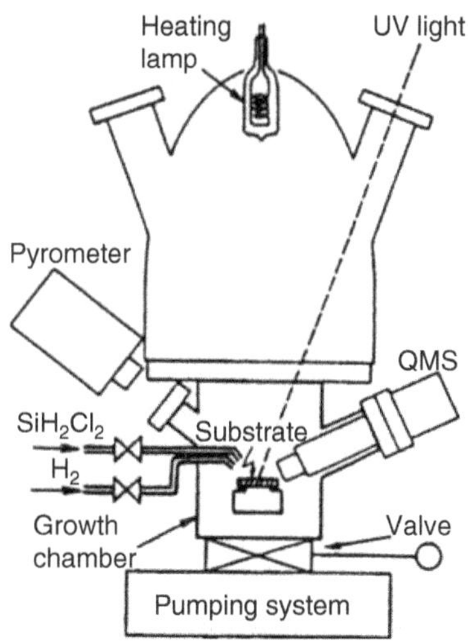

Figure 2.1 Silicon ALD experimental setup [2]. *Reprinted from J. Electrochem. Soc. 137, J. Niahizawa, K. Aoki, S. Suzuki, K. Kikuchi, Silicon Molecular Layer Epitaxy, 1898 © 1990, Reproduced by permission of ECS – The Electrochemical Society.*

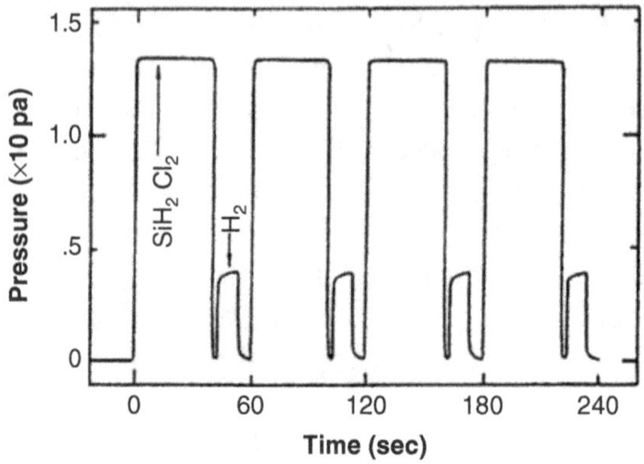

Figure 2.2 Pressure time plot for silicon deposition [2]. *Reprinted from J. Electrochem. Soc. 137, J. Niahizawa, K. Aoki, S. Suzuki, K. Kikuchi, Silicon Molecular Layer Epitaxy, 1898, © 1990, Reproduced by permission of ECS – The Electrochemical Society.*

pressure equal to 1 mtorr (0.133 Pa) and held there for 40 seconds. It is then evacuated in 3 seconds, at which point H_2 is injected at about 0.3 mtorr and held for 10 seconds, before being evacuated in 7 seconds. The total cycle time is 60 seconds.

Deposition is generally carried out by operating over a fixed number of cycles, and the deposition rate in Å/cycle can then be calculated once the film thickness is measured. Since we have a fairly good estimate of the thickness of a monolayer of silicon (1.4 Å for Si(001)), we can determine if the deposition rate is indeed a monolayer for each cycle. Secondly, we can determine whether or not the monolayer per cycle deposition is independent of deposition conditions. Figure 2.3 below [1] shows the deposition rate for the SiH_2Cl_2 + H_2 process as a function of the SiH_2Cl_2 pressure. Clearly, in the pressure range between 2×10^{-4} and 1×10^{-3} torr, the rate appears to be constant (825°C, B mode) and equal to a single monolayer of silicon per cycle. Also, similar behavior appears to occur for the temperature range between 815° and 850°C, although at 850°C the pressure range is from 1×10^{-4} to 3×10^{-4} torr.

Based on this evidence, it would seem clear that an ALD process has been established for these reactants and conditions. Interpretation of the experimental results requires more thought, however. First of all, the assumption was made that SiH_2Cl_2 was chemisorbing at the one monolayer level. However, it does not seem realistic to anticipate a single monolayer of silicon being produced by a monolayer of SiH_2Cl_2. This molecule is just too large to provide enough silicon atoms to create the observed monolayer

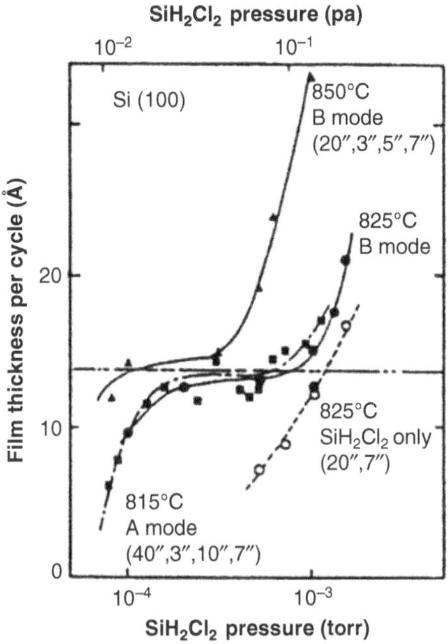

Figure 2.3 Film thickness per cycle as a function of pressure and temperature [1]. *Reprinted from J. Nishizawa, K. Aoki, S. Suzuki and K. Kikuchi, J. Cryst. Growth 99, J. Nishizawa, K. Aoki, S. Suzuki and K. Kikuchi, Molecular layer epitaxy of silicon, 502 © (1990), with permission from Elsevier.*

of silicon. Undoubtedly, at these high temperatures and long gas phase residence times, there must be gas phase reactions that are decomposing the SiH_2Cl_2 into smaller silicon bearing species. The authors [2] suggest that a monolayer of $SiCl_2$ is formed by dissociative chemisorption to allow the ALD growth to proceed.

If these high temperature results represent valid ALD depositions, then there should be a self-limiting monolayer of reactant initially on the substrate surface. This monolayer is more likely to be self-limiting if it is terminated either with a layer of hydrogen or a layer of chlorine, so that a second monolayer is less likely to adsorb. At these temperatures, any hydrogen on the surface will be desorbed. The question is whether or not any chlorine remains on the surface to passivate it. According to Yarmoff [3], their experiments show that when a surface is exposed to SiH_2Cl_2 at 800°C, and above, and then the chamber is evacuated, that no hydrogen or chlorine remains on the surface. These results suggest that if SiH_2Cl_2 decomposes into silicon atoms on hitting the hot substrate, then a self-limited monolayer should not be formed. Based on his data [3],

Yarmoff speculates that Nishizawa's ALD results [1, 2] may be the result of a dynamic process rather than a self-limited one.

Although this work appears to have shown, for the first time, ALD layer by layer growth of epitaxial silicon, regardless of the explanation, the growth temperature is much too high to be very useful. Another approach would be to reduce the first precursor with atomic hydrogen rather than the molecular species. This will be the topic of the next section.

2.1.1.2 *Using Atomic Hydrogen*

A number of researchers recognized that one way to reduce the required temperature for the ALD growth of epitaxial silicon was to use atomic hydrogen rather than H_2 as the reducer. Atomic hydrogen is extremely reactive, so it should promote reaction with a chemisorbed silicon-bearing compound (e.g. SiH_2Cl_2) at moderate temperatures [3–6]. In particular, such films have been deposited at temperatures as low as 550°C from SiH_2Cl_2 or some compound formed by its decomposition.

A schematic of the experimental reactor used to carry out this low temperature deposition is shown in Figure 2.4. In particular, we note

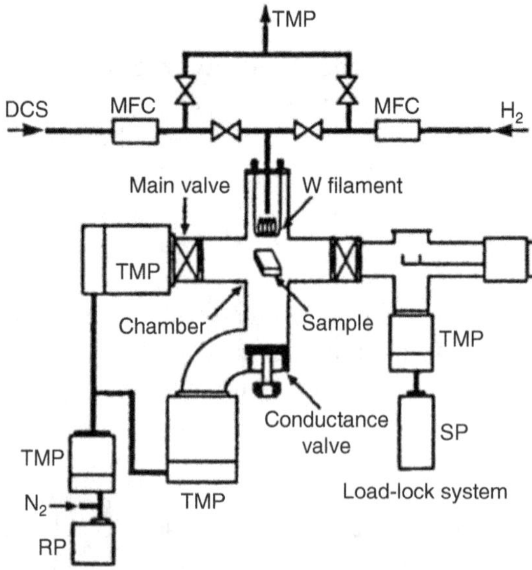

Figure 2.4 ALD reactor schematic [6]. *Reprinted with permission from E. Hasunuma, S. Sugahara, S. Hoshino, S. Imai, K. Ikeda and M. Matsumura, J. Vac. Sci. Technol A 16, 679 © (1998), American Vacuum Society.*

the heated tungsten filament used to generate the atomic hydrogen from hydrogen gas. The filament temperature was 2000°C, and the H atom exposure was maintained for 15 seconds. The system base pressure was 3×10^{-9} torr, and the sample was heated by passing an electric current through a tantalum sample holder.

The time cycle for this experiment is shown in Figure 2.5. Here we see that the SiH_2Cl_2 is introduced into the chamber, and held there for a period of time. Then, before hydrogen is introduced the SiH_2Cl_2 is pumped away. A short time later the hydrogen is introduced and the tungsten filament is turned on. Once the filament is turned off, the hydrogen continues to flow for a short time, and is then evacuated. A short time later, the SiH_2Cl_2 is introduced again to begin the second cycle.

Typical parameters for this process are shown in Table 2.1, where we can see that the cycle time is on the order of a minute, and the time the substrate is exposed to SiH_2Cl_2 is substantial (10 sec. in this case).

With the above apparatus and experimental conditions, these authors have demonstrated growth of single crystal silicon, one monolayer at a time, by an ALD process at moderate temperature. Such growth is illustrated in Figure 2.5, where the growth rate per cycle is plotted versus deposition temperature [6].

Referring to the discussion in Chapter 1 (section 1.3.5) we can argue that it is unlikely that we will be able to deposit silicon from

Table 2.1 ALD growth parameters [4]. *Reprinted from Thin Solid Films 225, S. Imai, T. Iizuka, O. Sugiura and M. Matsumura, Atomic layer epitaxy of Si using atomic H, 168 © (1993), with permission from Elsevier.*

Duration of adsorption interval	10 s
Duration of exhaust interval	30 s
Duration of reduction interval	15 s
Duration of filament cooling interval	10 s
Duration of exhaust interval	30 s
Substrate temperature	650°
SiH_2Cl_2 flow rate	2.6 standard $cm^3\ min^{-1}$
H_2 flow rate	1.0 standard $cm^3\ min^{-1}$
Filament temperature	< 2000°C
Substrate orientation	(111)
Substrate size	5 mm × 28 mm

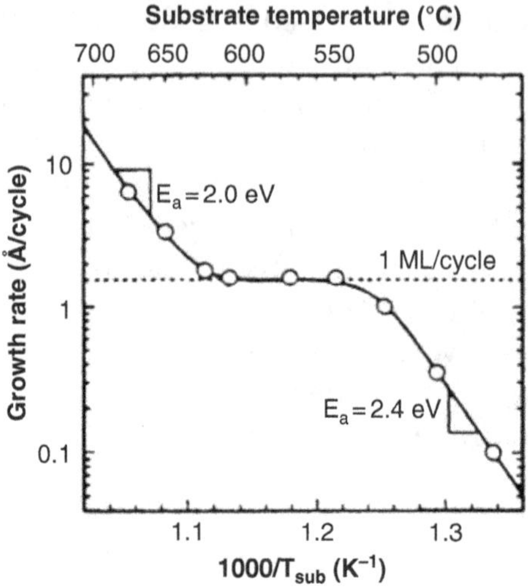

Figure 2.5 ALD growth of epitaxial silicon from SiH_2Cl_2 and H [6]. *Reprinted with permission from E. Hasunuma, S. Sugahara, S. Hoshino, S. Imai, K. Ikeda and M. Matsumura, J. Vac. Sci. Technol A 16, 679 © (1998), American Vacuum Society.*

SiH_2Cl_2 at a rate of one ML/cycle. Even if we assume that SiH_2Cl_2 will be cracked by heat from the heated substrate [8] to yield $SiCl_2$ and H_2, we still find it difficult to argue for a one ML/cycle deposition of silicon from $SiCl_2$. There is one other possibility. If SiH_2Cl_2 decomposes into SiClH, this molecule could chemisorb onto the silicon surface in such a way as to permit a one ML/cycle growth. A possible scenario for such a deposition is shown in Figure 2.6.

At (a) we have the clean silicon surface with its dangling bonds. After the SiH_2Cl_2 exposure, and SiHCl is formed, the SiHCl adsorbs dissociatively on the surface until the surface is covered 50% with chlorine and 50% with H, as shown in (b). With enough heat, the adsorbed H will desorb, leaving additional free sites for SiHCl to adsorb onto (c). Eventually, all the H will desorb, and we will have a silicon surface that is fully passivated by chlorine atoms (d). Then an exposure to hydrogen atoms in the next step will create HCl molecules that will be driven off by the same heat that drove off the H atoms, thereby removing the chlorine. The end result is another clean silicon surface ready for the growth of another monolayer.

There may also be another possible explanation for the experimental results observed. Evidence for some gas phase reaction allowing one ML/cycle deposition is shown in Figure 2.7. Short

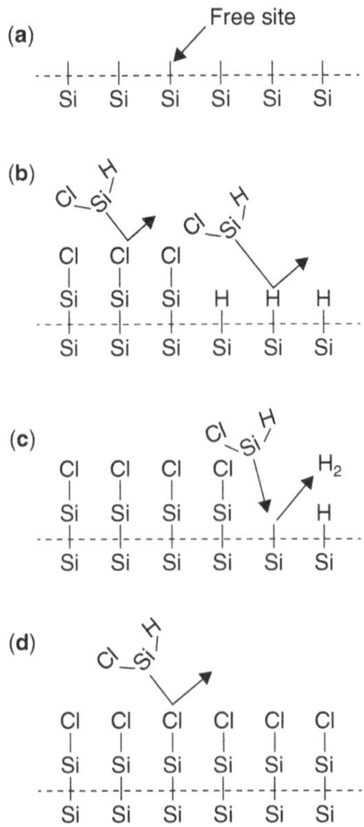

Figure 2.6 Postulated surface reaction sequence [6]. *Reprinted with permission from E. Hasunuma, S. Sugahara, S. Hoshino, S. Imai, K. Ikeda and M. Matsumura, J. Vac. Sci. Technol A 16, 679 © (1998), American Vacuum Society.*

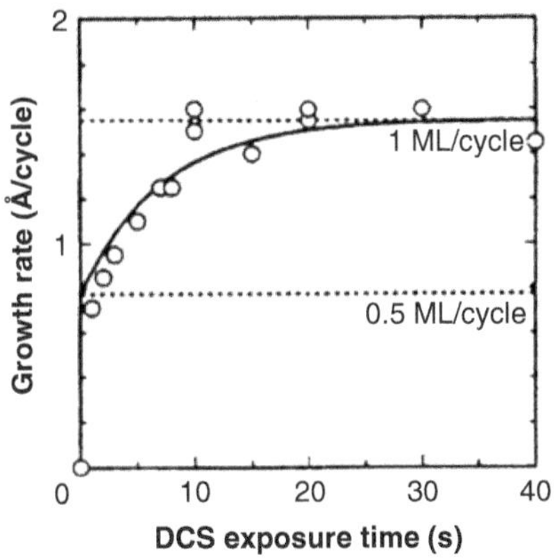

Figure 2.7 Growth rate of silicon under high pressure conditions [6]. *Reprinted with permission from E. Hasunuma, S. Sugahara, S. Hoshino, S. Imai, K. Ikeda and M. Matsumura, J. Vac. Sci. Technol A 16, 679 © (1998), American Vacuum Society.*

DCS time exposures produce 0.5 ML/cycle growth. As the exposure time increases, the growth rate becomes the desired one ML/cycle [6].

2.1.2 Other Processes

There are many silicon bearing compounds that could be tried as precursors for epitaxial silicon growth. A few have been studied and have shown promise, because they have been shown to be able to form the desired self-limiting monolayer [9–13].

2.1.2.1 Chlorosilanes

Hexachlorodisilane (Si_2Cl_6) has been evaluated at temperatures in the 400°– 525°C range with atomic hydrogen being used as the second reactant [9]. Evidently, Si_2Cl_6 chemisorbs onto silicon or germanium, in a self-limiting fashion, by decomposing into two $SiCl_3$ molecules. Once a complete monolayer is formed it is terminated with chlorine atoms, and no further adsorption occurs. Then when the surface is exposed to a flux of atomic hydrogen (produced by a heated tungsten filament), the chlorine is removed. At the temperatures used, any hydrogen on the surface is rapidly desorbed. The timing sequence was as follows:

First a Si_2Cl_6 exposure for 1.8 L, then 2 minute pump out, then an H_{at} exposure at 1.5 L, followed by a 4 minute pump out.

Although only 20 cycles were run, it appeared that silicon growth was occurring linearly on a germanium substrate. Due to the large volume of the original $SiCl_3$ molecules that are believed to be completely covering the surface, the silicon growth rate was found to be 1/3 ML/cycle.

Detailed measurements of deposition using trichlorosilane have also been made [14]. The dominant surface reactions have been found to be

$$SiHCl_3 \rightarrow {}^*SiCl_2 + HCl \tag{2.3}$$

and

$$^*SiCl_2 + H_2 \rightarrow Si + 2HCl \tag{2.4}$$

where $^*SiCl_2$ represents the chemisorbed species

2.1.2.2 *Silanes*

Another novel approach to epitaxial silicon growth by ALD is based on the self-limited monolayer chemisorption of Si_2H_6, and the subsequent removal of surface hydrogen. There have been two removal methods; using a He plasma [10] or by temperature cycling [15]. The use of disilane is motivated by the desire to remove any possibility of residual chlorine remaining in the deposited film. Once the disilane is chemisorbed to a silicon surface, we end up with a new silicon surface which is terminated with hydrogen atoms. The problem then is how to remove these hydrogen atoms so that the next chemisorption cycle can proceed. One way to do it would be to simply chemisorb onto a heated surface, hot enough to desorb the hydrogen. However, this high temperature (~600°C) would also decompose any disilane hitting the surface immediately, and cause a CVD type of deposition rather than ALD. Use of energetic He atoms to remove the hydrogen allows the process to be run at a lower temperature and the temperature cycling method allows the disilane adsorption to take place at a temperature where it is stable followed by evacuation and rapid cycling to a higher temperature where hydrogen desorption can take place without danger of CVD.

The experimental setup used to explore the approach using He plasma is shown in Figure 2.8. As can be seen, it is a UHV chamber

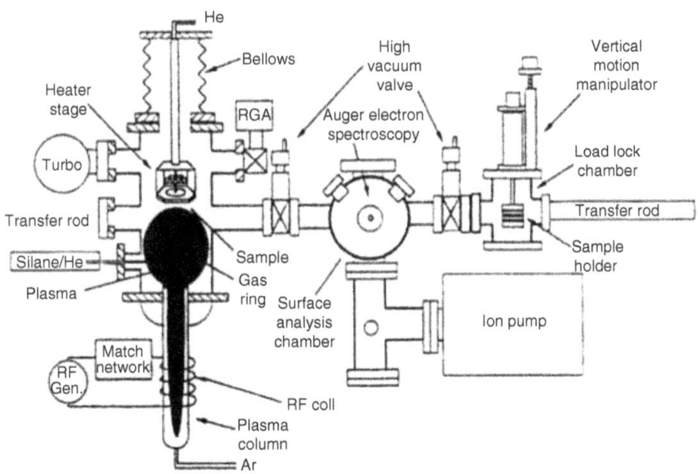

Figure 2.8 ALD system schematic with He plasma source [10]. *Reprinted with permission from Thin Solid Films 225, A. Mahajan, J. Irby, D. Kinosky, R. Qian, S. Thomas, S. Banerjee, A. Tasch and T. Picraux, Si atomic layer epitaxy based on Si2H6 and remote He plasma bombardment, 177 © (1993), American Vacuum Society.*

equipped with an RF plasma induction coil operating at 13.56 MHz, a load lock, and a surface analysis chamber. System base pressure is reported to be 3×10^{-9} torr.

The procedure developed included a remote hydrogen plasma to clean the initial substrate surface. This hydrogen terminated original silicon surface was then treated with a flux of excited He atoms and/or He ions from the He discharge, to remove any adsorbed hydrogen atoms. This should have yielded the original silicon surface with dangling bonds and free of hydrogen atoms. Next, this virgin surface was dosed with disilane which chemisorbed, and created an adsorbed layer of new silicon atoms covered with hydrogen atoms. Finally, this surface hydrogen was removed with the He flux again, and a fresh silicon surface was recovered. This cycle was repeated as many times as necessary to grow the desired silicon film thickness.

The wafer temperature was left at 400°C so as not to allow decomposition of the disilane on impact. The choice of the proper power level and gas pressure for the He plasma bombardment was not trivial. It was found that too high a pressure permitted effective hydrogen removal only in a very narrow range of discharge powers. Their qualitative results of discharge power range (dark bars) vs. gas pressure are shown in Figure 2.9.

For the process they ran, the r.f. power chosen was 30 watts at a pressure of 100 mtorr. The disilane doses were in the range of 10^5–10^6L, and produced growth rates of about 0.15 ML/cycle.

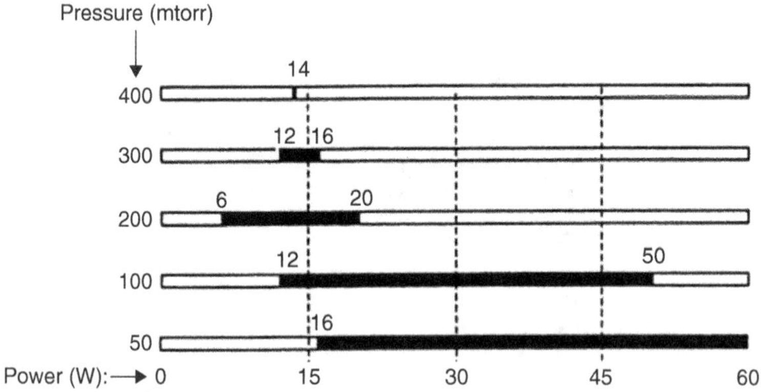

Figure 2.9 Power vs. pressure with a He discharge at 250°C [10]. *Reprinted with permission from Thin Solid Films 225, A. Mahajan, J. Irby, D. Kinosky, R. Qian, S. Thomas, S. Banerjee, A. Tasch and T. Picraux, Si atomic layer epitaxy based on Si2H6 and remote He plasma bombardment, 177 © (1993), American Vacuum Society.*

The He bombardment was continued for 2–3 min. per cycle. Although good quality epitaxial silicon was grown, by an ALD process, the growth rate was less than the desired 1 ML/cycle.

The demonstration that energetic He atoms can remove adsorbed hydrogen at low temperatures is significant. Calculations also confirmed that approximately 3 surface H atoms can be displaced by each He atom in the energy range of 15–60 ev, with no bulk silicon atoms displaced during this exposure.

When one of the higher silanes which chemisorbs in a self-limiting monolayer (e.g. disilane, trisilane), is used as the silicon precursor, it will decompose on the surface into either SiH_2 or SiH_3 molecules. In either case, the surface will be terminated by hydrogen atoms. If the surface is heated to above approximately 500°C, hydrogen can be desorbed, so if the precursor will chemisorb the desired monolayer at a lower temperature, we can achieve silicon epitaxy by thermally cycling the wafer. Such a process has been demonstrated with the use of disilane [15] and trisilane [11]. With trisilane, the low temperature was set at 344°C, when the trisilane was adsorbing, and the high temperature was 555°C which desorbed any hydrogen left on the silicon surface. The total cycle time was 2 min. The high and low temperatures were varied to confirm the ALD growth mode. Growth rate curves are shown in Figure 2.10. With disilane, the respective temperatures were 400°C (low) and 550°C (high). A plot of the temperature cycling profile is shown in Figure 2.11.

In this case, the film growth rate saturated at ~0.5 Å/cycle equivalent to 0.3 monolayers/cycle. The growth rate was also stable over the (low) deposition temperature range from 350°C to 450°C. Above 500°C it increased rapidly because of CVD reactions.

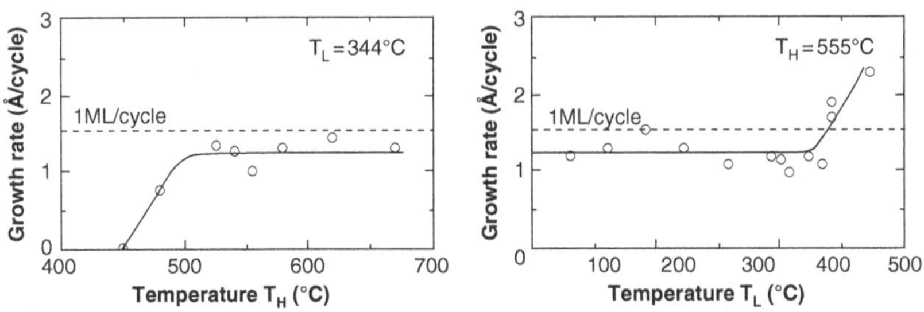

Figure 2.10 Thickness per cycle vs. temperature [11]. © *1991 The Japan Society of Applied Physics.*

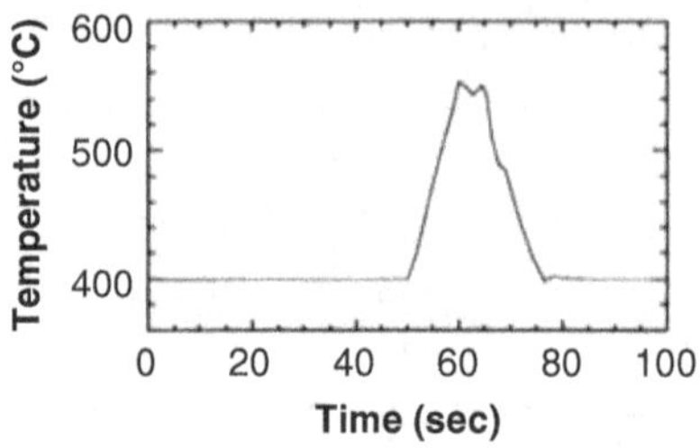

Figure 2.11 Temperature cycling profile [15] *Reprinted from J. Cryst. Growth 209, J. Nishizawa, A. Murai, T. Ohizumi, T. Kurabayashi, K. Ohtsuka, T. Yoshida, Observation and control of surface reaction during Si molecular layer growth, 327 © (2000), with permission from Elsevier.*

Although single crystal silicon has been grown by these methods, the film quality was not quite as good as desired, and further research needs to be done on this approach.

2.1.3 Epitaxial Germanium

Atomic layer epitaxy deposition of pure germanium films has been carried out using dimethylgermane ($(CH_3)_2GeH_2$ or DMGe) as the germanium precursor [12, 13, 16] As will be seen for many other organo-metallic precursors in later chapters, this molecular species chemisorbs onto the germanium surface with a self-limiting monolayer. As before, this is the first requirement for a successful ALD process. The second step involves removing all species from this layer other than Ge, and then proceeding to a second layer of DMGe. One approach would be to simply use heat to drive off the methyl radical and the hydrogen. A second approach would be to use atomic hydrogen to convert the CH_3 radicals to CH_4 that would then leave as a stable gas and leave Ge behind. In fact, both approaches have been shown to work [12], although the second [13] appears to have a wider temperature window. One advantage of the atomic hydrogen process is that the overall cycle time can be shortened since the exposure to the H removes the excess methyl and hydrogen terminating surface species faster than be purely thermal desorption.

In the experimental apparatus used to deposit Ge by the thermal process [12], particular care was taken to avoid carbon contamination from the roughing pump. The base pressure was 3×10^{-9} torr in the main chamber. For the thermal process, a typical set of conditions was 0.3 sec. for the adsorption step, 100 sec. for the desorption step, a pressure of 29 mtorr for the DMGe and a temperature of 522°C. ALD growth was achievable , however the temperature window was very narrow (510–530°C). At the same time, no carbon or oxygen contamination was found in the Ge films, at least to the precision of Auger analysis (~ 1%), and they appeared to be epitaxial germanium.

Typical operating conditions for the atomic hydrogen experiments are summarized below. The Ge substrate is held at a lower temperature (e.g. 480°C), and the total cycle time is approximately 100 sec. The adsorption phase of the DMGe involves a fairly high vapor pressure (29 mtorr), but a short time exposure (0.3 sec). This amounts to an exposure of 8.7×10^3 L, which should ensure the availability of enough molecules to form the desired monolayer of the precursor. The hydrogen atom source was operated for 10 sec. after which the hydrogen was evacuated. The growth rate per cycle was 1.4 Å, that is 1 ML/cycle, over a range of DMGe exposure times from 0.3 to 3 s .The growth rate per cycle is shown in Figure 2.12 as function of H_{at} exposure time, for the above conditions.

In Figure 2.12 we see that for H_{at} exposure times > 10 sec. the deposition rate is again one ML/cycle. For H_{at} exposure times of

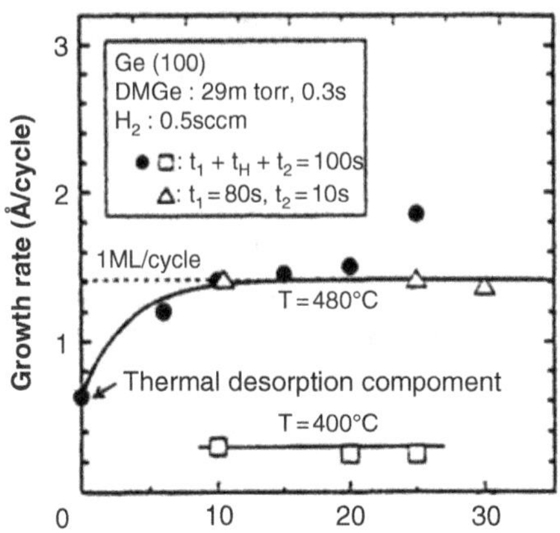

Figure 2.12 Growth per cycle vs. H_{at} exposure time [13]. *Reprinted from Appl. Surf. Sci 90, S. Sugahara, M Kadoshima, T. Kitamura, S. Imai and M. Matsumura, Atomic hydrogen-assisted ALE of germanium, 349 © (1995).*

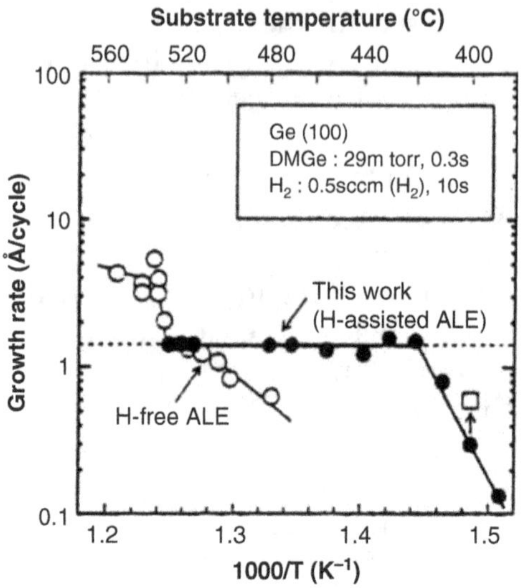

Figure 2.13 Growth per cycle vs. temperature [13]. *Reprinted from Appl- Surf. Sci 90, S. Sugahara, M Kadoshima, T. Kitamura, S. Imai and M. Matsumura, Atomic hydrogen-assisted ALE of germanium, 349 © (1995).*

less than 10 sec, there is not enough H_{at} flux to reduce all of the DMGe precursor, and the growth rate is less than one ML/cycle.

In Figure 2.13 we show the growth rate per cycle as a function of substrate temperature for the H_{at} assisted case. Also in this plot is shown the growth rate for the thermal process. As can be seen, both the process temperature is higher, and the temperature window is much narrower when thermal desorption is used, vs. when H_{at} are used.

Another curious feature of Figure 2.13, however, needs to be explained. Here the dark circles represent the H_{at} process. Since hydrogen atoms are so reactive, the drop in deposition rate as temperature is reduced is not what we would have expected. After eliminating several possible explanations for this anomalous behavior, the authors concluded that H_2 was adsorbing on the clean Ge surface, and the surface temperature had fallen below the temperature needed for its desorption. To verify this conjecture, they ran an experiment at 400°C, but introduced a 10 second temperature pulse of 480°C just after completion of the H_{at} exposure. They found that the growth rate increased from 0.3 to 0.6 Å/cycle with this modification. In Figure 2.13 this data point is represented by the open square.

References

1. J. Nishizawa, K. Aoki, S. Suzuki and K. Kikuchi, *J. Cryst.Growth* 99, 502 (1990).
2. J. Nishizawa, K. Aoki, S. Suzuki and K. Kikuchi, *J. Electrochem. Soc.* 137, 1898 (1990).
3. Y.A. Yarmoff, D.K. Shuh, T.D. Durbin, C.W. Lo, D.A. Lapiano- Smith, F.R. McFeely and F.J. Himpsel, *J. Vac. Sci. Technol. A10*, 2303 (1992).
4. S. Imai, T. Iizuka, O. Sugiura and M. Matsumura, *Thin Solid Films* 225, 168 (1993).
5. S. Sugahara, E. Hasunuma, S. Imai and M. Matsumura, *Appl. Surf. Sci.* 107, 161 (1996).
6. E. Hasunuma, S. Sugahara, S. Hoshino, S. Imai, K. Ikeda and M. Matsumura, *J. Vac. Sci. Technol A* 16, 679 (1998).
7. D.D. Koleski and S.M. Gates, *J. Appl. Phys.* 76, 1615 (1994).
8. C. Sasaoka and A. Usui, *Appl. Surf. Sci.* 82/83, 348 (1994).
9. D.D. Koleski and S.M. Gates, *Appl. Phys. Lett.* 64, 884 (1994).
10. A. Mahajan, J. Irby, D. Kinosky, R. Qian, S. Thomas, S. Banerjee, A. Tasch and T. Picraux, *Thin Solid Films* 225, 177 (1993).
11. S. Imai, S. Takagi, O. Sugiura and M. Matsumura, *Jpn. J.Appl.Phys.* 30, 3646 (1991).
12. S. Sugahara, T. Kitamura, S. Imai and M. Matsumura, *Appl. Surf. Sci.* 82/83, 380 (1994).
13. S. Sugahara, M Kadoshima, T. Kitamura, S. Imai and M. Matsumura, *Appl. Surf. Sci.* 90, 349 (1995).
14. H. Habuka, Y Aoyama, S. Akiyama, T. Otsuka, W. Qu, M. Shimada and K. Okuyama, *J. Cryst. Growth* 207, 77 (1999).
15. J. Nishizawa, A. Murai, T. Ohizumi, T. Kurabayashi, K. Ohtsuka and T. Yoshida, *J. Crys. Growth*, 209, 327 (2000).
16. S. Sugahara and M. Matsumura, *Appl. Surf. Sci.* 112, 176 (1997).

3

III-V Semiconductor Films

In addition to the single element semiconductor materials reviewed in the last chapter, there has been a great deal of interest in depositing compound semiconductors by ALD. The initial aim was to deposit high quality epitaxial films of exceptional purity. In addition to this property, compound semiconductors, as compared to elemental semiconductors, must also have perfect stoichiometry. Since the process is even more complex than for silicon and germanium, achieving the above goals is a considerable challenge. In the present chapter we will focus on III-V semiconductors. More recently there has been a body of work directed towards photonic structures using semiconductors where polycrystalline material has been deposited on regular array structures such as inverse opals.

3.1 Gallium Arsenide

Studies of the deposition of GaAs from an organometallic or halogen bearing Ga precursor, and an arsenic bearing compound have been numerous. Under the appropriate operating conditions, alternate exposures of the Ga precursor and the arsenic precursor deposits epitaxial GaAs films a monolayer at a time. However, the mechanism by which this ALD deposition is achieved is still not completely clear in spite of numerous surface chemistry studies. We will review a number of these studies, and their sometimes conflicting results.

One of the difficulties in carrying out this process by strictly thermal means with organometallic precursors, is that in some cases precursors may decompose or desorb if the temperature is high enough to obtain epitaxy. This limits the choice of precursors. Or, alternately, the precursors may not react rapidly enough if the temperature is too low. For this reason additional studies have been done using hydrogen radical enhancement, and halogen based precursors as well, in an attempt to broaden the "temperature window". Both of these areas will be treated in later sections.

3.1.1 Organometallic Precursors

Of the many studies that have been done of GaAs deposition by ALD using organometallic precursors, most have involved thermal reactions between adsorbed trimethylgallium (TMGa, $Ga(CH_3)_3$) and adsorbed arsine (AsH_3), so we will review these first [1–18]. There is no question that alternate exposures of TMGa and arsine to a heated surface will enable epitaxial GaAs to be grown a monolayer at a time. Some of these studies were done in UHV chambers with each precursor being exposed to the substrate and then being evacuated with a vacuum pump [1, 5–7, 10, 12, 13, 15]. Others have been done in flow systems where each precursor is removed with a purge gas [2–4, 8, 9, 11, 14, 16–18]. In the flow systems, operating pressures have varied from atmospheric to 20 torr depending on the research group. Although all of these groups reported ALD, there were significant differences in their results. Since this particular ALD process is especially sensitive to the surface chemistry of the two precursors, and their reaction either near to or on the substrate surface, this is probably the reason for the different results reported.

We will review the results obtained in the vacuum evacuated systems first. In a vacuum chamber similar to the one shown in Figure 3.2 [1], a pressure cycle with an overall cycle time of 30 sec was used, and is illustrated below in Figure 3.1. For this study depositions were done on GaAs(100) substrates. Arsine was held in the chamber for 20 sec, then the chamber was evacuated for 3 sec, then TMGa was introduced and held for 4 sec, and again the chamber was evacuated for 3 sec.

As evidence that a self-limiting monolayer could be grown, Figure 3.2 shows the film thickness per cycle vs. TMGa pressure [6]. For this data, the solid symbols represent results for the timing cycle of Figure 3.1. The timing for the data shown by the open triangles was (20 s, 1 s, 4 s, 1 s). As can be seen, for temperatures in the 450–500°C range, the deposition rate is approximately 1 ML/cycle, since the thickness of a single monolayer of GaAs would be 2.83 Å, provided there is enough TMGa to saturate the surface. A similar behavior was found with the arsine exposure which also saturated.

A number of interesting features were observed. First, the films were selective. That is, Si_3N_4 was used as a mask, and no deposition was observed on this material.

Secondly, the films obtained were observed to be p-type with a carrier density of 10^{18}–10^{20} cm^{-3}, which may be indicative of carbon contamination from the TMGa, or a lack of perfect stochiometry. A similar chemistry has been studied in an atmospheric pressure reactor with a spinning susceptor [2–4, 14, 16]. A schematic of this reactor is shown below in Figure 3.3 [3]. In this reactor design, the substrate sits on a silicon carbide coated graphite susceptor which rotates underneath a similar fixed part with two openings. Above each opening is located a gas injection port, with one used to introduce TMGa and the other used to introduce arsine. A hydrogen

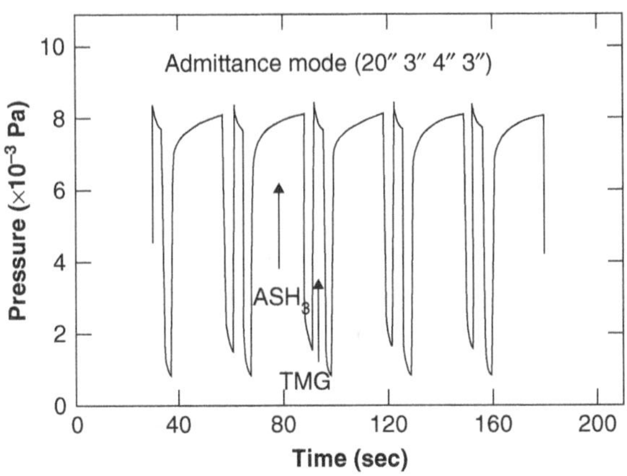

Figure 3.1 Chamber pressure vs. time for ALD of GaAs [1]. *Reprinted from J. Electrochem. Soc. 132, J-I. Nishizawa, H. Abe, T. Kurabayashi, Molecular Layer Epitaxy, 1197 © 1985, Reproduced by permission of ECS – The Electrochemical Society.*

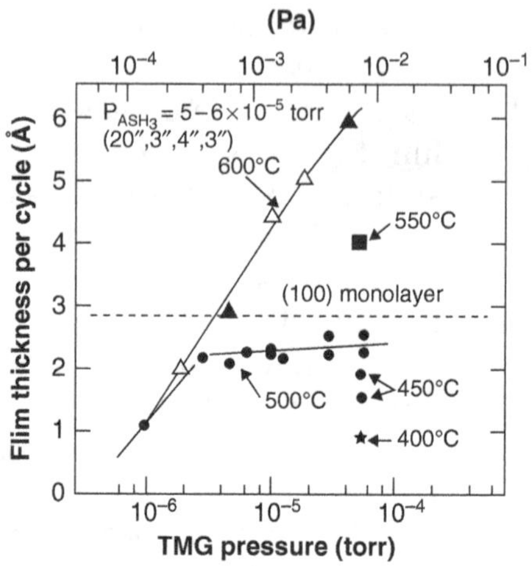

Figure 3.2 Film thickness per cycle vs. TMGa pressure [6]. *Reprinted from J. Electrochem. Soc. 134, J-I. Nishizawa, T. Kurabayashi, H. Abe, N. Sakurai, Deposition Mechanism of GaAs Epitaxy,945* © 1987, Reproduced by permission of ECS – The Electrochemical Society.

flow is placed between these two reacting gas streams to keep them from intermixing. Each reacting gas stream only impinges on the substrate for a short period of time. The small spacing between the two parts of the susceptor causes the reacting gas to be sheared off before the substrate approaches the second reactant flow. One revolution constitutes one deposition cycle. A typical time for one cycle was 2.6 s, and the exposure time to each gas was on the order of 0.3 s. The gallium precursor was TMGa, delivered from a hydrogen bubbler with a hydrogen flow between 0.75 to 10 sccm. The arsenic precursor was arsine (5% in hydrogen) and its flow was 25 sccm. Both precursors were delivered in a hydrogen carrier gas at 500 sccm flow rates, at a pressure of one atmosphere. Single crystal GaAs was grown by this technique as evidenced by transmission electron diffraction patterns [4]. The growth per cycle was about one monolayer, and was independent of TMGa flow, as shown in Figure 3.4. Similarly, growth rates were independent of arsine flow and substrate temperature within the range of 450–700°C. Depositions were carried out on (100) and (111)B GaAs crystal orientations.

Finally, these authors reported room temperature carrier concentrations and Hall mobilities at several deposition temperatures, and these are shown in Table 3.1 [3]. The data from the first 630°C run

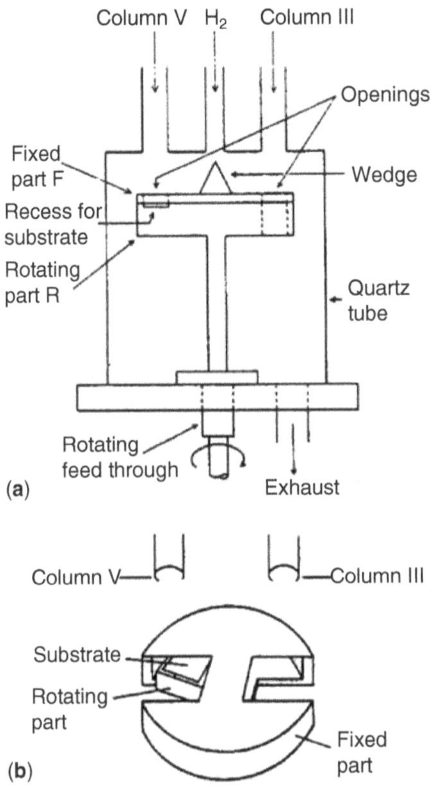

Figure 3.3 Atmospheric pressure ALD reactor with rotating susceptor [3]. *Reprinted from J. Cryst. Growth 77, M.A. Tischler and S.M. Bedair, Growth and characterization of compound semiconductors by atomic layer epitaxy, 89 © (1986) with permission of Elsevier.*

(a) was taken with a new susceptor, and showed what appears to be a very clean film. The second data point (b) was taken after many runs, and may have reflected a worn silicon carbide coating on the carbon susceptor leading to carbon contamination.

Apparently, the GaAs epitaxial films grown by this technique could be grown over a much wider temperature range, and with less carbon contamination than other techniques [16]. One explanation could be that the substrate is exposed to a high flux of reactants (atmospheric pressure) for a very short time, compared to the high vacuum reactor systems.

Another concept, which uses a high flux of reactants used a vertical MOCVD reactor operating at pressures from 15 to 20 torr [9]. A variation in the reactor operation, as compared to traditional MOCVD, was

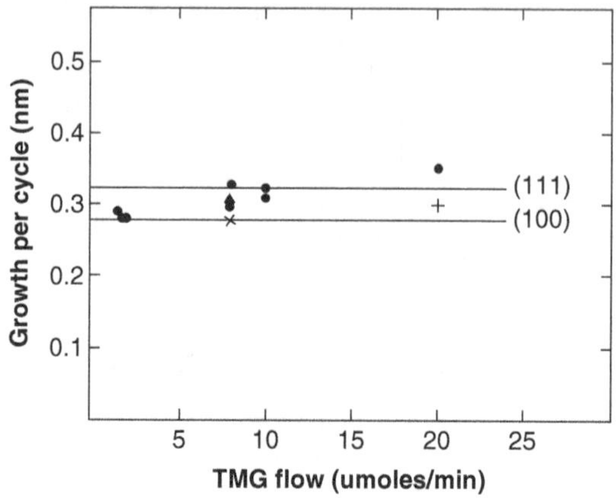

Figure 3.4 Thickness per cycle of ALD GaAs vs TMGa flux [4]. *Reprinted with permission from M.A. Tischler and S. M. Bedair, Appl. Phys. Lett. 48, 1681 © (1986), American Institute of Physics.*

Table 3.1 Room temperature carrier concentration and Hall mobility for GaAs grown by atomic layer epitaxy [3]. *Reprinted from J. Cryst. Growth 77, M.A. Tischler and S.M. Bedair, Growth and characterization of compound semiconductors by atomic layer epitaxy, 89 © (1986) with permission of Elsevier.*

Growth temperature (°C)	Carrier concentration P (cm^{-3})	Hall mobility (cm^2/V.s)
450	2×10^{16}	215
500	5×10^{17}	190
630a	1×10^{15}	280
630b	2×10^{17}	210

that a nozzle was added through which the precursors could be introduced. Since the wafer was held at temperatures in the range of 500°C, allowing the reactants to approach it slowly produced unwanted gas phase reactions and a CVD like growth rate with temperature. This prevented layer by layer growth.. By injecting the gases at high velocity, the residence time above the wafer was reduced and so were the gas phase reactions. This enabled the desired growth of a monolayer per cycle. The alternating pulses of TMGa and AsH_3 were separated by hydrogen purges. Typically, the hydrogen purges were run for 3 seconds, and the arsine pulse was 10 seconds.

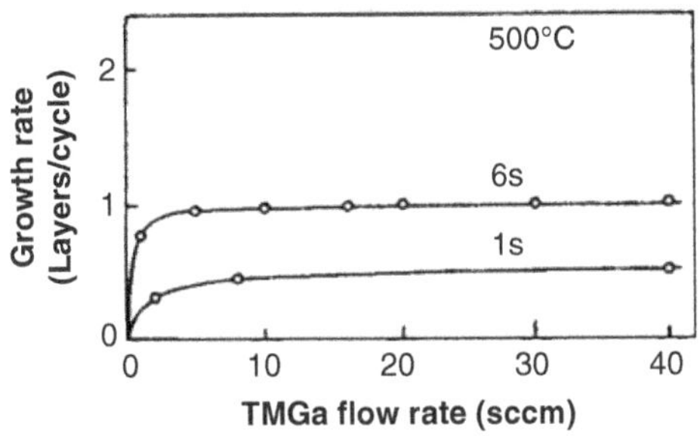

Figure 3.5 Growth rates of GaAs as a function of TMGa flow rate for two TMGa pulse durations [9]. *Reprinted from M. Ozeki, K. Mochizuki, N. Ohtsuka and K. Kodama, Appl. Phys.Lett. 53, 1509 © (1988), American Institute of Physics .*

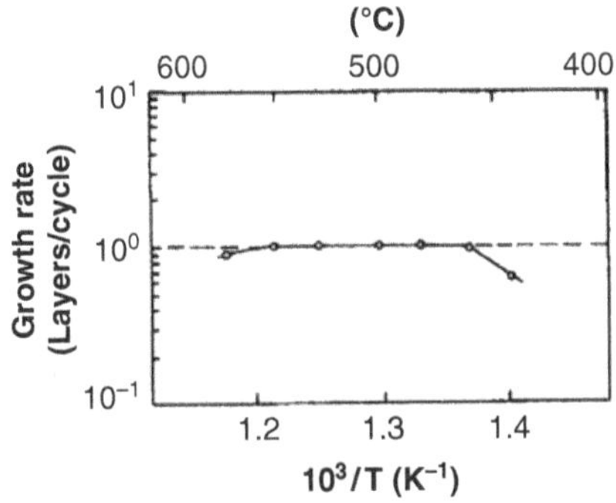

Figure 3.6 Growth rate of GaAs as a function of temperature [9]. *Reprinted from M. Ozeki, K. Mochizuki, N. Ohtsuka and K. Kodama, Appl. Phys.Lett. 53, 1509 © (1988), American Institute of Physics.*

As seen in Figure 3.5, the growth rate exhibits the desired value of one monolayer per cycle for TMGa flow rates greater than 5 sccm, as long as the TMGa pulse length is greater than 6 seconds. The growth rate as a function of temperature is shown in Figure 3.6, where the TMGa pulse length was 15 seconds. Here we see that the temperature window for ALD growth ranges from 460 to 550°C. This compares to the temperature window found in the atmospheric pressure reactor [4], which was reported to be 450 to 700°C. It is also interesting to note that the growth rate per cycle drops when the temperature exceeds 550°C. This is in contrast to other ALD systems where at elevated temperatures the deposition rate per cycle goes up (e.g. Chapter 2, Figure 3.10). As was discussed in Chapter 1 (see section 1.3.3), the drop is probably caused

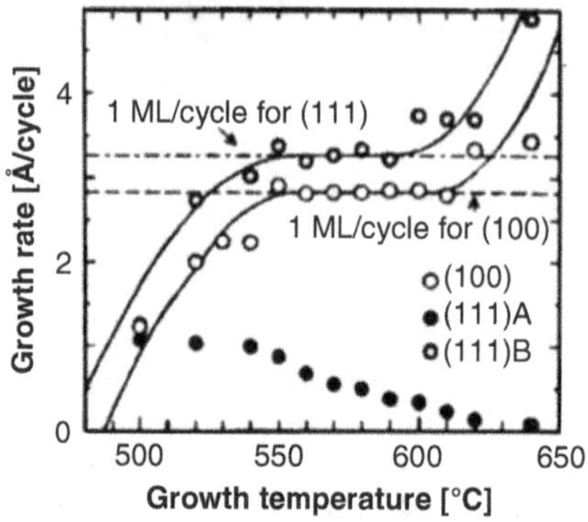

Figure 3.7 GaAs growth rates, from TMGa and arsine, as a function of temperature for three GaAs crystal orientations [17]. Reprinted from *Appl. Surf. Sci. 103, J-S. Lee, S. Iwai, H. Isshiki, T. Meguro, T. Sugano and Y. Aoyagi, Self-limiting growth on nominally oriented (111)A GaAs substrates in atomic layer epitaxy, 275 © (1996), with permission from Elsevier.*

by the desorption of a stable molecule, whereas the increase could be due to decomposition of the precursor. Finally, these GaAs films were found to have n-type conductivity, with a carrier concentration of 2×10^{14} cm^{-3} and a mobility of 6.5×10^4 cm^2/V s at 77°K.

In another study using an MOCVD reactor [19], the ALD growth of GaAs is studied when nitrogen is used as the carrier gas rather than hydrogen. The reactor did not employ the jet inlet described above, and was operated at 70 torr with TMGa and arsine. The temperature window is slightly broadened from 490–500°C with hydrogen to 490–520°C with nitrogen. In both higher vacuum systems [13] and higher pressure (~10 torr) MOCVD systems [17], it has been observed that GaAs will grow in a self-limiting fashion on (100) and (111)B GaAs crystal faces, but very low growth rates per cycle are seen on the (111)A face.

An illustration of this behavior is illustrated in Figure 3.7. The explanation for this must lie in the fact that the (111)B face is terminated with arsenic atoms, and the (111)A face is terminated by gallium atoms. Also, the (100) face is terminated by both gallium and arsenic. Then if TMGa adsorbs onto an arsenic atom, but will not adsorb onto a gallium atom, this would explain the above behavior [13].

In addition to the extensive research on thermal ALD of GaAs from TMGa and arsine described above, there has also been some study of the use of other precursors. Of particular interest has been the use of triethylgallium (TEGa) rather than TMGa and arsine, because of some indications from MOCVD studies that better quality films can be realized with these precursors [5]. Although there was some indication that GaAs could be grown by ALD in a vacuum system from these precursors [7] at 320°C on the (100) face of GaAs, the temperature window was very small. In another experiment in an atmospheric pressure MOCVD type reactor [19], however, it was shown that ALD films could be grown from these precursors within a temperature range of 340–350°C. The growth rate increased with temperature over the range 300 to 400 °C with a very narrow temperature window at ~350 °C where the growth rate was 1 ML/cycle .When the temperature was held at 350°C, the growth rate seemed independent of arsine flow over a wide range. It was independent of TEGa flow over a more limited range but continued to increase for higher flows.

The fact that substantially different results were obtained in an atmospheric pressure reactor [19] as compared to a vacuum reactor [7], again points out that for these complex ALD systems the type of reactor used, and the operating conditions employed can have a substantial effect on the results obtained.

Although other organometallic precursors have been considered, only a few have been examined in any detail. Of particular interest is a study of the replacement of arsine with tris-dimethylamino-arsenic (TDMAAS, $As(N(CH_3)_2)_3$) and its use with TEGa [20]. The depositions were carried out in a vacuum chamber with a 1 second pulse of TEGa, a 10 second evacuation, and a 1 second pulse of TDMAAS followed by an 18 second evacuation. In this case, the temperature window is expanded significantly to 440°C–490°C.

The proposed explanation for this improvement in the width of the temperature window hinges on the desorption characteristics of methyl (CH_3), ethyl (C_2H_5) and amine (NCH_3) radicals. The argument postulated, is that when TMGa chemisorbs to an arsenic terminated surface, it may decompose so that there may be MMGa ($GaCH_3$), DMGA ($Ga(CH_3)_2$) or TMGa molecules on the surface. Then the methyl radical will terminate the surface, and since the desorption time for the methyl radical from a gallium surface is reported to be [20] 50 seconds at 485°C, the methyl radicals would inhibit the further chemisorption of TMGA. This should limit the gallium supply to approximately one monolayer. Then the arsenic bearing molecule (e.g. AsH_3) will react with this monolayer forming a monolayer of GaAs with the remaining methyl radicals and hydrogen being desorbed. On the other hand, when TEGa is chemisorbed at moderately high temperatures it may decompose into Ga and all of the ethyl radicals may be rapidly desorbed since the desorption time for ethyl radicals [20] is reported to be less than 0.1 seconds at 470°C. Therefore, additional TEGa molecules will continue to arrive depositing more gallium, and more than a monolayer is formed on each pulse.

When TDMAAS is used as the arsenic precursor, there is a source of amine radicals that can remain on the surface and migrate to any gallium atoms formed, thereby limiting the further adsorption of TEGa, so that a self-limiting condition is easier to achieve. It is reported that the desorption time for the amine radical is about 3 seconds in the temperature range of these experiments [20].

Other than TMGa and TEGa, the only other organometallic gallium precursor reportedly used for the ALD of GaAs is trisneopentylgallium (TNPGa) [18]. For depositions at a growth temperature of 470°C both the TMGa and TNPGa precursors demonstrate a one monolayer deposition per cycle when the exposure time is greater than 5 seconds, as shown in Figure 3.8. The reactor was operated at 50 torr, and the substrates were GaAs (100). The arsine precursor was tertiarybutylarsine (TBA).

Based on all of the above studies it is clear that thermal ALD of GaAs is possible. However, the appropriate process conditions, as well as the film quality will depend to a large degree on the reactor type and process used. As well, the precise mechanisms for

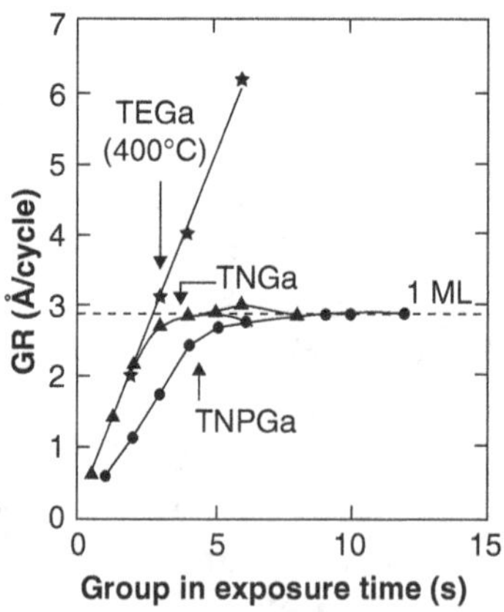

Figure 3.8 Growth rate per cycle vs. exposure time for TMGa,TEGa and TNPGa precursors [18]. *Reprinted from R. Ares, S.P. Watkins, P. Yeo, G.A. Horley, P. O'Brien and A.C. Jones, J. Appl. Phys. 83, 3390 © (1998), American Instutute of Physics.*

achieving monolayer per cycle growth are not well understood, and may be different for different reactors and/or processes [21].

Finally, it should be noted that there have been many attempts to enhance the ALD process when organometallic precursors were used for GaAs deposition [1, 5, 7]. Most of these efforts used optical radiation (i.e. lasers or mercury lamps) to stimulate the reaction between TMGa or TEGa and arsine, in hopes of broadening the "temperature window" for the particular process. Although they achieved some success [5, 7] optical enhancement techniques are difficult to apply to commercial reactors, so they will not be reviewed further. However, there is one study which used hydrogen atoms to stimulate the reaction between TMGa and arsine [22], and they appeared to have been successful in obtaining a wider temperature window. This system employed a plasma generated by a 2.45 GHz microwave source through which hydrogen flowed to create a flux of hydrogen atoms. The reactor was quartz, and operated at 20 torr. The plasma generator was operated continuously, and either H_2 or He was passed through it. The flux of hydrogen atoms was terminated when the He flow replaced the H_2 flow. The TMGa and arsine were diluted in H_2, and pure H_2 was used as a purge gas to remove the TMGa vapor or arsine from the chamber after each pulse. The timing cycle was

$$AsH_3/H_2/TMGa/H_2/H/H_2$$

and a typical set of values, in seconds, was

$$2/15/\text{variable}/20/3/3$$

ALD growth seems to have been achieved at temperatures from 400–500°C. The lower value is somewhat below temperatures reported earlier for the strictly thermal process. However, the film grown at 400°C levels off at a growth rate per cycle of less than one monolayer, suggesting that although a "self limiting" growth is obtained there is not enough thermal energy available to cause a true monolayer to form in each cycle. In addition, these authors report that they achieved depositions at temperatures as low as room temperature. However, the film quality was poor (bad morphology), and mirror smooth GaAs layers were only seen when the temperature rose to 400°C. Apparently, all the films were Ga:As (1:1), with the only observable impurity being oxygen as evaluated by Auger analysis. Also, the higher the temperature, the lower the oxygen contamination.

More recently, quantum mechanical studies of the surface reactions which take place when ALD of GaAs takes place on silicon substrates have been carried out using computational methods aiming to clarify the deposition conditions favourable for growth [23].

3.1.2 Halogen Precursors

In addition to the organometallic Ga precursors discussed in the previous section it is also possible to deposit GaAs films from the Ga halides [24–26]. The difficulty is that there are no convenient liquid halide precursors with reasonable vapor pressures available in the desirable molecular form. For example, as will be discussed, GaAs can be grown from GaCl and arsine. However, the stable forms of gallium chloride are Ga_2Cl_4 and $GaCl_3$ and they are solids at room temperature. The reason that the GaCl molecule is so desirable is that we wish to grow a single crystal material with each growth cycle. In order to do this we need a full coverage of the surface with Ga atoms. If there is more than one chlorine atom attached to each gallium atom, then we will not have sufficient gallium to form the single crystal structure. So the question becomes, how do we create a flux of a gallium halide with each molecule only having one halogen atom? One approach is to pass an acid vapor over a sample of solid gallium at a high temperature. For example, HCl vapor in a hydrogen carrier gas is reported to produce a vapor of GaCl when passed over gallium at 780°C [24]. Similar results are reported for HBr and HI to produce GaBr and GaI [26]. Finally, in one study the compound diethylgalliumchloride (DEGaCl) has been reported to decompose into GaCl and hydrocarbons [27] when heated. A simple atmospheric pressure reactor used to carry out ALD was described by Usui and Sunakawa [24] using a two-chamber system where GaCl and As gases flowed in their own process tube [24]. In this reactor the exposure to each reactant was changed by physically moving the substrate back and forth between two chambers. Similar results were obtained in reactors that used hydrogen purges between exposures [22, 25].

In all of these studies it has been shown that the growth of one monolayer of GaAs is possible, by the ALD technique. It is largely independent of the flux of the gallium halide or its exposure time. There is, however, a dependence on substrate temperature. Growth rates fall below one monolayer per cycle as the temperature increases above 425°C for the iodine system, and above 450°C for the bromide system [26]. At least up to 600°C the growth of one monolayer per cycle holds for the chloride case. A final note of interest is that the growth of GaAs is selective. That is, it is reported to not grow on SiO_2 while growing on GaAs substrates.

In all of these studies, it is argued that a monolayer of the gallium halide chemisorbs onto the arsenic atoms available on the substrate surface, and that the halide will not chemisorb onto itself. Therefore, a self limiting monolayer is formed. This monolayer is then reduced to a layer of GaAs by exposure to AsH_3. Presumably, the arsine provides hydrogen which combines with the halide atoms and the resulting acid vapor (i.e. HCl) can desorb. At the same time a layer of arsenic is left behind so that the process can be repeated.

Although it is clear from these studies that single crystal GaAs can be grown at temperatures as low as 400°C by these halide processes, there is no information reported on the chemical composition of these films. Even trace quantities of impurities such as carbon or chlorine could be important.

3.2 Other III-V Semiconductor Films

Although most studies reported of ALD growth of III-V materials has been of GaAs films, other III-V compound semiconductors have also been deposited. Specifically, there have been reports of studies of deposition of InAs [27], AlAs [28, 29], GaN [30, 31], GaP [33, 38], InGaN [34], InGaAs [35] and AlGaAs [36]. Reactors and procedures are similar to those described earlier in this chapter, except that different reactants needed to provide the aluminum, indium, nitrogen and phosphorous atoms were chosen. These have been mainly organometallics. To deposit materials containing aluminum, trimethylamine-alane (TMAAl), trimethylaluminum (TMA), and triethylaluminum (TEA) were used. The indium precursors employed were ethyldimethylindium (EDMIn), trimethylindium (TMIn), and triethylindium (TEIn). For growth of nitrides (i.e. GaN) ammonia was used, and for phosphides (i.e. GaP) either phosphine (PH_3) or phosphorous vapor was selected. There has also been some work using $GaCl_3$ as a precursor for GaN which showed a deposition rate of around 0.5 monolayer/cycle, presumable due to steric hindrance by the precursor molecules [37].

Much of the interest in these additional III-V semiconductors occurred because of the desire to create superlattices or superalloys, where new and novel structures composed of alternating layers of different materials can be grown. Particularly, if the individual layers can be deposited controllably in thicknesses of several monolayers, unique materials can be created which can have new and interesting properties.

3.3 Applications

3.3.1 Photonic Structures

Recently, with the growth of interest in photonics where tunable luminescence and reflectance properties can be obtained, compounds such as GaAs, GaN, GaP and AlGaN have all been studied for photonic structures. The structures which have been investigated are inverse opals, where a template of a regular array of nanospheres has been conformally coated by ALD and the nanospheres then dissolved away leaving a hollow honeycomb-like structure [38–40] or a series of quantum dots obtained by adjusting the deposition parameters so that the ALD layer was not conformal but consisted of isolated nanoparticles by Stranski- Krastanov growth [41, 42]. Quantum wire structures using localized ALD growth have also been reported [43–45].

3.3.2 Transistors

The use of ALD to produce doped compound semiconductors and their use in transistors has been reported [46–49]. Diethyl zinc (DEZ $(C_2H_5)_2Zn$) and diethyl selenium (DES $(C_2H_5)_2Se$) have been used as p- and n-type dopants respectively and p-type doping with carbon has also been achieved by residual carbon remaining from incomplete decomposition of the precursor being retained in the deposited film. A complex stack of doped and undoped layers was thus obtained as shown in Figure 3.9.

n^{++} Se doped 2×10^{19} cm^{-3} 300 Å
n^{+} Se doped 5×10^{18} cm^{-3} 1000 Å
undoped 300 Å
p^{+} C doped 5×10^{18} cm^{-3} 36 Å
undoped 1500 Å
n^{+} Se doped 5×10^{18} cm^{-3} 500 Å
Si doped GaAs (100) substrate

Figure 3.9 Layer stack produced by ALD in transistor structure [48].

References

1. J-I. Nishizawa, H. Abe and T. Kurabayashi, J. Electrochem. Soc. 132, 1197 (1985).
2. S.M. Bedair, M.A. Tischler, T. Katsuyama and N.A. El-Masry, *Appl. Phys. Lett.*47, 51 (1985).
3. M.A. Tischler and S.M. Bedair, *J. Cryst. Growth* 77, 89 (1986).
4. M.A. Tischler and S. M. Bedair, *Appl. Phys. Lett.* 48, 1681 (1986).
5. J-I. Nishizawa, H. Abe, T. Kurabayashi and N. Sakurai, *J. Vac. Sci. Technol. A4*, 706 (1986).
6. J-I. Nishizawa, T. Kurabayashi, H. Abe and N. Sakurai, *J. Electrochem. Soc.* 134, 945 (1987).
7. J-I. Nishizawa, T. Kurabayashi, H. Abe and N. Sakurai, *J. Vac. Sci. Technol. A5*, 1572 (1987).
8. P.D. Dapkus and S.P. DenBaars, Proc. Intern. Electron Devices Mtg. IEEE 472 (1988).
9. M. Ozeki, K. Mochizuki, N. Ohtsuka and K. Kodama, *Appl. Phys.Lett.* 53, 1509 (1988).
10. M.L. Yu, U. Memmert and T.F. Kuech, *Appl. Phys. Lett.* 55, 1011 (1998).
11. Y. Sakuma, M. Ozeki, N. Ohtsuka and K. Kodama, *J. Appl. Phys.* 68, 5660 (1990).
12. A. Watanabe, T. Kamijoh, T. Isu and Y. Katayama, *Vacuum* 41, 965 (1990).
13. J. Nishizawa, *J. Cryst. Growth* 115, 12 (1991).
14. K.G. Reid, U.M. Urdianyk and S.M. Bedair, *Appl. Phys. Lett.* 59, 2397 (1991).
15. B.Y. Maa and P.D. Dapkus,. *Lett Appl. Phys.* 58, 1762 (1991).
16. M. Bedair, *J. Vac. Sci. Technol. B* 12, 179 (1994).
17. J-S. Lee, S. Iwai, H. Isshiki, T. Meguro, T. Sugano and Y. Aoyagi, *Appl. Surf. Sci.* 103, 275 (1996).
18. R. Ares, S.P. Watkins, P. Yeo, G.A. Horley, P. O'Brien and A.C. Jones, *J. Appl. Phys.* 83, 3390 (1998).
19. H. Yokoyama, M. Shinohara and N. Inoue, *Appl. Phys. Lett.* 59, 2148 (1991).
20. K. Fujii, I. Suemune and M. Yamanishi, *Appl. Phys. Lett.* 61, 2577 (1992).
21. J.R. Creighton and B.A. Banse, *MRS Proceedings*, (Eds. T.F. Kuech, P.D. Dapkus and Y. Aoyagi) 222, 15 (1991).
22. M. de Keijser and C. van Opdorp, *Appl. Phys. Lett.* 58, 1187 (1991).
23. L.Dong, Q-Q.Sun, Y. Shi, H. Liu, C. Wang,S-J. Ding and D. W. Zhang, *Appl. Phys. Letts.* 92, 111105 (2008).
24. A. Usui and H. Sunakawa, *Jpn. J. Appl. Phys.* 25, L212 (1986).
25. K. Nishi, A. Usui and H. Sakaki, *Appl. Phys. Lett.* 61, 31 (1992).
26. T. Taki and A. Koukitu, *Appl. Surf. Sci.* 112, 127 (1997).
27. M. L. Yu, *J. Appl. Phys.* 73, 716 (1993).

28. N. Ohtsuka, K. Kitahara, M. Ozeki and K. Kodama, *J. Crystal Growth* 99, 346 (1990).
29. K. Fujii, Y. Suemune and M. Yaminishi, *Appl. Phys. Lett.* 62, 1420 (1993).
30. M. Asif Khan, R.A. Skogman, J.M. Van Hove, D.T Olson and J.N. Kuznia, *Appl. Phys. Lett.* 60, 1366 (1992).
31. M. Asif Kahn, J.N. Kuznia, D.T. Olson, T. George and W.T. Pike, *Appl. Phys. Lett.* 63, 3470 (1993).
32. Y. Sakuma, K. Kodama and M. Ozeki, *Appl. Phys. Lett.* 56, 827 (1990).
33. T. Taki, T. Nakajima, A. Koukitu and H. Seki, *J. Crystal Growth* 183, 75 (1998).
34. E.L. Piner, M.K. Behbehani, N.A. El-Masry, F.G. McIntosh, J.C. Roberts, K.S. Boutros and S.M. Bedair, *Appl. Phys. Lett.* 70, 461 (1997).
35. N. Ohtsuka, K. Kodama, M. Ozeki and Y. Sakuma, *J. Crystal Growth* 115, 460 (1991).
36. J.R. Gong, D. Jung, N.A. El-Masry and S.M. Bedair, Appl. Phys. Lett. 57, 400 (1990).
37. O. H. Kim, D. Kim and T. Anderson, *J. Vac. Sc. Technol. A* 27, 923 (2009).
38. I. M. Povey, D. Whitehead, K. Thomas, M. E. Pemble and M. Bardosova, *Appl. Phys. Letts.* 89, 104102 (2006).
39. I.M. Povey, M. Bardosova, F.C. Dillon, F. Chalvet, M.E. Pemble and K. Thomas, *Thin Solid Films* 517, 811 (2008).
40. E. Graugnard, V. Chawla, D. Lorang and C. J. Summers, *Appl. Phys. Letts* 89, 211102 (2006).
41. J.H. Kim, Y.J. Park, Y.M. Park, J.D. Song, J.I. Lee and T.W. Kim, *Appl. Surf. Sci.* 253, 3503 (2007).
42. Y-L. Tsai, J-R. Gong, T-Y. Lin, H-Y. Lin, Y-F. Chen and K-M. Lin, *Appl. Surf. Sci.* 252, 3454 (2006).
43. H. Isshiki, Y. Aoyagi and T. Sugano, *Appl. Surf. Sci.* 112, 122 (1997).
44. H. Isshiki, S. Iwai, T. Meguro, Y. Aoyagi and T. Sugano, *J. Cryst. Growth* 145, 976 (1994).
45. H. Isshiki, S. Iwai, T. Meguro, Y. Aoyagi and T. Sugano, *J. Appl. Phys.* 78, 7277 (1995).
46. T. Kurabayashia, H. Kikuchia, T. Hamanoa and J. Nishizawa *J. Cryst. Growth* 229 , 147 (2001).
47. J. Nishizawa and T. Kurabayashi, *Thin Solid Films* 367, 13 (2000).
48. J. Nishizawa and T. Kurabayashi, *Adv. Mater. Optics Electron.*, 7, 183 (1997).
49. R. Hayes, R. Bhat, E. Colas and R. Esagui, *IEEE Trans. Electron Dev.* 35, 2433 (1988).

4

Oxide films

4.1 Introduction

Metal oxides are difficult to deposit by thermal CVD techniques at moderate temperatures. Since ALD methods allow the separation of each reactant, until they react at the substrate surface, more highly reactive species may be used (without any danger of gas phase reactions causing gas phase nucleation) and lower process temperatures are possible. The precursor chemistry and the substrate structure determine the beginning of the growth of the film. Depending on the surface energy of the substrate or the already grown film, the film continues to grow as amorphous or crystalline. Some materials like Al_2O_3 require high thermal energy to grow as crystalline and therefore in ALD deposition temperatures it grows generally as amorphous. In many materials the crystallinity appears in the ALD range at a very early stage of growth and is dependent on the film thickness. Depending on the surface energies crystal phase and orientation usually transform when the film grows in thickness. It is very important to understand the exact thin film structure, morphology, composition, and electrical properties to be able to connect them with the characteristics and the behavior of the material.

ALD has found interesting and great potential in new industrial applications, such as diffusion barriers on polymer substrates. This also raise demand to develop the efficient low temperature ALD processes. Many oxides can be deposited by thermal ALD at moderate temperatures. However, there are limitations as to how low a temperature can be used, because the chemical kinetics of the surface reaction may be too slow. Taking advantage of the energetic radicals created in a plasma, not only can this limitation be minimized, but use can be made of energetic ion bombardment to modify the growing films structure, if desired.

The charateristics of a precursor are demanding; high reactivity, volatility, and good thermal stability over a wide temperature range are desired. Since there are many commercially important applications for such films, a great deal of research and development has been done on their deposition by ALD. In this chapter we will attempt to highlight some of this work.

4.2 Aluminum Oxide

Aluminum oxide represents many ways a good example in ALD, especially in practical applications. That is, it is well studied especially by one of its best processes TMA and water [1], and TMA and oxygen plasma [2]. These recent review articles will provide the reader the complete source of aluminum oxide processes studied. It can be deposited at relatively wide temperature window starting from room temperature allowing coatings also on temperature sensitive substrates. Many of the current industrial applications, potential future applications, as well as early applications, are based on aluminum oxide. These are presented more detailed in Chapter 8. Aluminum oxide thin films can be grown by ALD from a variety of aluminum source molecules and several oxidizers. The films can be amorphous polycrystalline or single crystal depending on the deposition temperature. As will be shown, in general the deposition temperatures are low, and the lowest temperatures are achieved with the most reactive oxidizers [1–16].

All of the ALD studies are predicated on the ability to deposit a monolayer of an aluminum precursor in a self-limiting fashion. That is, a molecular species containing aluminum adsorbs on a substrate, and after one monolayer has formed, additional molecular species do not attach to the surface. Depending on the surface temperature, and the molecular species chosen, the molecules of the aluminum containing species may undergo some chemical decomposition (i.e. dissociative chemisorption). However, such decomposition is not a necessary condition to achieve ALD growth.

In general, less than one monolayer of aluminum oxide is formed at each ALD cycle. A simple explanation of this phenomenon can be offered based on the "steric" effect. That is, the TMA molecule is quite large, and it is therefore not possible to position enough aluminum atoms on the surface in a single monolayer of TMA to yield a full monolayer of aluminum oxide for each cycle.

A simple calculation can illustrate this point. For Al_2O_3 grown from TMA we can estimate the deposition rate for an ALD process.

Aluminum atoms available in one monolayer of TMA ($Al(CH_3)_3$)
TMA M.W. = 72, ρ = 0.752 gms/cm^3
Molecular mass = $72 \times 1.67 \times 10^{-24} = 1.2 \times 10^{-22}$ gms
Molecules/vol = 0.752 gms/cm^3/1.2×10^{-22} gms = 6.27×10^{21} #/cm^3
Molecules/area = $(6.27 \times 10^{21})^{2/3} = 3.4 \times 10^{14}$ #/cm^2

Therefore, since there is only one Al atom per TMA molecule, there are only 3.4×10^{14} aluminum atoms available in each initial TMA monolayer.

Aluminum atoms in one monolayer of aluminum oxide
Al_2O_3 M.W. = 102, ρ = 3.97 gms/cm^3
Molecular mass = $102 \times 1.67 \times 10^{-24} = 1.7 \times 10^{-22}$ gms
Molecules/vol = 3.97 gms/cm^3/1.7×10^{-22} gms = 2.3×10^{22} #/cm^3
Molecules/area = $(2.3 \times 10^{22})^{2/3} = 8 \times 10^{14}$ #/cm^2

In order for one monolayer of Al_2O_3 to be formed, there will have to be double this number of aluminum atoms on the surface.
Aluminum atoms/area = $2 \times 8 \times 10^{14} = 1.6 \times 10^{15}$ #/cm^2

However, we only have 3.4×10^{14} aluminum atoms per cm^2 available from a single TMA monolayer. Therefore, we can only grow, at the most, 3.4/16 = 0.2 monolayers of Al_2O_3 per cycle.

Although this is a smaller number than reported, some of the difference may be attributable to the chemical reactions that can occur when TMA chemisorbs onto a surface [8]. In other words, the monolayer may well be formed of smaller molecules than TMA.

For more fundamental examination a theoretical model in literature is available, which describes the ALD growth per cycle based on chemisorption and surface saturation, and also discusses the resulting crystallinity of the film as function of ALD process parameters [1, 17–19]. The gas-solid reactions that can result the attachment of the compound reactant onto a solid surface are classified as ligand exchange, dissociation and association. After chemisorption of a compound reactant on a solid surface, the surface saturation with adsorbed species is then limited by steric hindrance of the ligands and by the limited number of bonding sites on the surface.

4.2.1 Processes and Properties of Aluminum Oxide

4.2.1.1 Chlorine Processes

One of the earliest studies to demonstrate the ALD growth of aluminum oxide [3] showed that single crystal material can be grown from aluminum chloride ($AlCl_3$) vapor and oxygen gas at 500°C on niobium and at 600°C, and above, on sapphire. Also, the same film grown on sapphire at 500°C was polycrystalline, and when it was grown at 400°C it was amorphous. The overall reaction is presumed to be

$$4AlCl_3 + 3O_2 \rightarrow 2Al_2O_3 + 6Cl_2 \quad (4.1)$$

The depositions were carried out in a vacuum chamber, with a turbopump capable of exhausting the reaction chamber to a pressure of 6×10^{-09} torr. The $AlCl_3$ vapor was held in the chamber for 30 seconds at pressures of $3.75–7.5 \times 10^{-05}$ torr, and the oxygen was introduced as an 85% He - 15% O_2 mixture for 60 seconds at a pressure of $6.0{\times}10^{-05}$ torr. The $AlCl_3$ exposure amounted to approximately 1,500 L (L= 1×10^{-06} torr-sec).

Using "reflection electron diffraction" these authors confirmed that the films grown at 600°C on single crystal sapphire were indeed homoepitaxial Al_2O_3 or sapphire. Deposition rates were observed to be ~ 0.9Å/cycle, and the refractive index was 1.76, which is very close to that of bulk sapphire. For deposition cycles up to 500, the desired linear deposition rate is illustrated in Figure 4.1.

These authors did not, however, report any chemical analysis of their films. Therefore, although the film was likely close to stochiometric composition there was no data on possible impurities such as chlorine.

As demonstrated above, $AlCl_3$ clearly adsorbs/chemisorbs a single monolayer per pulse, and this enables the deposition of Al_2O_3

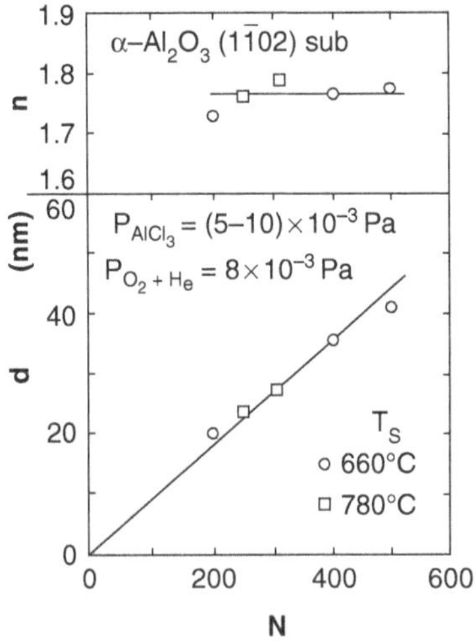

Figure 4.1 Refractive index and film thickness as a function of number of deposition cycles [3]. *Reprinted with permission from Appl. Phys. Lett. 51, G. Oya, M. Yoshida and Y. Sawada, 1143, © 1987, American Institute of Physics.*

by ALD with molecular oxygen as the oxidizer. In later studies [9], a similar process was run successfully with water vapor as the oxidizing agent. Again, the overall reaction should be

$$2AlCl_3 + 3H_2O \rightarrow Al_2O_3 + 6HCl \tag{4.2}$$

Here a mechanism that depends on OH radicals terminating the alumina surface has been proposed, thereby preventing any other $AlCl_3$ molecules from adsorbing, so that a self-limiting monolayer will form.

Although OH may play some role in the $AlCl_3$ plus H_2O process, the earlier paper [3] using O_2 demonstrates that the presence of OH radicals is not a requirement for the formation of a self limiting monolayer of $AlCl_3$. Of course, what does form the monolayer may be some mixture of $AlCl_3$, $AlCl_2$, or AlCl depending on the process temperature.

In addition to $AlCl_3$ as an aluminum precursor, another study explored the use of dimethyl aluminum chloride or DMACl ($Al(CH_3)_2Cl$) and water as reactants [10]. ALD films were grown over the temperature range of 125–500°C. The impurities observed in the films were chlorine, carbon, and hydrogen. Their concentrations depended on the temperature of deposition, as shown in

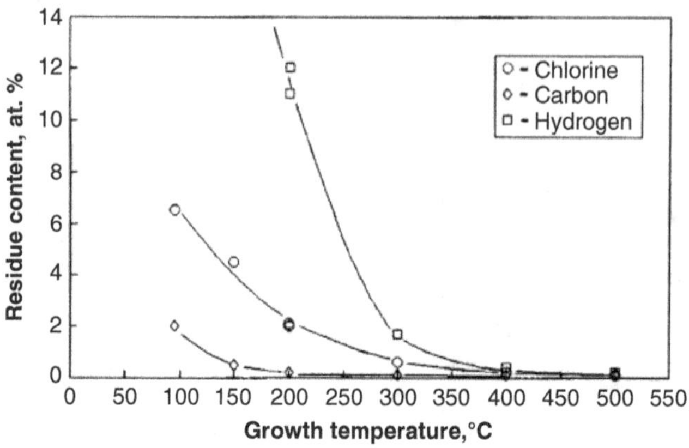

Figure 4.2 Impurity content of aluminum oxide films deposited by ALD [10]. *Reprinted with permission from J. Vac. Sci. Technol. A 15, K. Kukli, M. Ritala and M. Leskelä, 2214, © 1997, American Vacuum Society.*

Figure 4.2. It is interesting to observe that there was reported to be approximately 60% hydrogen in films grown at 95°C, and that this concentration only reduced to 12% at 200°C. Also, the chlorine concentration is still approximately 0.5% at a 300°C deposition temperature.

Although a successful ALD process can be demonstrated with a chlorine based precursor (i.e. $AlCl_3$), there are three concerns that occur when using it. One, is that it does not have a very high vapor pressure (it is a solid at room temperature with a melting point of 190°C), and so it is more difficult to introduce into the chamber.

Secondly, chlorine based reaction products (i.e. HCl) will arise when oxidants other than oxygen are used, and these will be very corrosive to the reactor and pumps. Finally, there is concern, in general, when even a fraction of a percent of chlorine is left in the grown film. In fact, it has been shown that when Al_2O_3 is grown from $AlCl_3$ and water vapor [9] even at 500°C there is approximately 0.5% chlorine in the film.

4.2.1.2 Organometallic Precursor Processes

For these reasons, additional studies have explored the use of aluminum organometallic precursors. A number of these studies have looked at the use of TMA (trimethylaluminum) with water as the oxygen source [4, 8, 9, 11–15, 23–29]. TMA is a liquid at room temperature, and has a vapor pressure similar to that of water.

Therefore, they are both quite easy to introduce into a vacuum chamber using short time doses. The presumed reaction will be

$$2Al(CH_3)_3 + 3H_2O \rightarrow Al_2O_3 + 6CH_4 \tag{4.3}$$

It is reported that films were grown at temperatures of room temperature up to 500°C. Here it must be realized that TMA starts to decompose at around 300°C, and deposition above this temperature is not likely to follow the ALD mode. That is, deposition is outside the ALD window.

Later studies have shown that the overall deposition rate below 300°C still varies with deposition temperature. Reported values for TMA and water processes are in the range of 0.9 to 1.3 Å/cycle [23, 28, 29]. Typically in modern commercial ALD reactors the cycle time varies from less than 1 second to several seconds depending for example on deposition temperature and surface area of the substrates. In one study the growth per cycle of 0.9 Å/cycle was achieved at deposition temperature of 220°C with total cycle length of 1.5 seconds [29]. It should be noted that approximately one half a monolayer of Al_2O_3 is being formed during each cycle.

As far as chemistry of the film is concerned, the films are found to be stoichiometric Al_2O_3 by RBS analysis and free of carbon contamination, at least to less than 0.5%. The refractive index is reported to be between 1.65 and 1.70 (which is somewhat less than the value of 1.76 for single crystal sapphire), and the film is amorphous by TEM.

Electrical behavior of the dielectric was evaluated by carrying out capacitance vs. voltage measurements using aluminum dots on silicon. The dielectric constant was 7, which is somewhat low compared to sapphire. Resistivity was ~ 10^{17} ohm-cm, and breakdown fields of ~ (6–8) × 10^6 V/cm were reported.

In an attempt to lower the deposition temperature of the above process, hydrogen peroxide vapor can be substituted for water vapor as the oxidizer [5–7]. Films of quality similar to those deposited with water vapor were obtained at 150°C. Growth rate was with hydrogen peroxide was somewhat higher.

Interestingly, it has been shown for the H_2O_2 process [7] that the deposition rate of aluminum oxide is essentially independent of temperature in the wide temperature range of 140–750°C, as shown in Figure 4.3. Also, as can be seen, ALD growth using N_2O doesn't occur at any temperature.

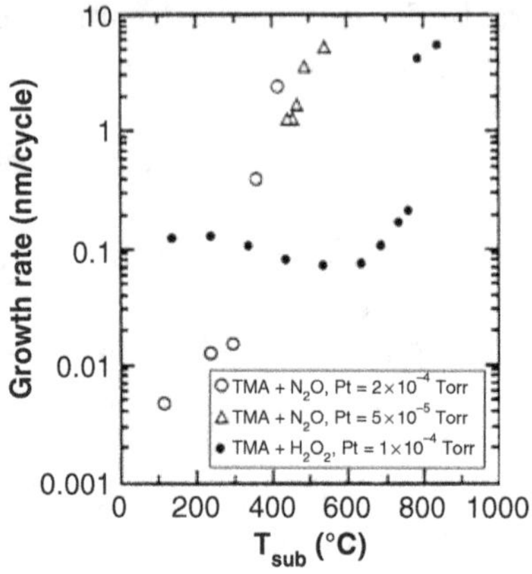

Figure 4.3 Growth rate of Al_2O_3 from TMA as a function of temperature [7]. *© 1993. The Japan Society of Applied Physics.*

As in the earlier study [4], the film formed from H_2O_2 is reported to be stoichiometric by RBS analysis. XPS analysis confirms that the oxygen atoms in the film are bound to aluminum atoms. The amorphous character of the films was confirmed by RHEED analysis. The dielectric constant, the refractive index, the electrical resistivity, and the breakdown strength were all similar to the earlier study with water vapor.

It was also interesting to observe that when films grown at room temperature were annealed at 150°C the original thickness of 2270Å reduced to 1200Å. Since RBS analysis would not reveal the presence of hydrogen, the room temperature films were either low density or contained a substantial quantity of hydrogen or both. However, the properties of the annealed films were similar to those originally deposited at 150°C.

Another oxidizer in addition to water or hydrogen peroxide is ozone, and a number of papers have explored films grown from TMA by this technique [16–19, 30] For this case, the overall reaction would look like

$$2Al(CH_3)_3 + O_3 \rightarrow Al_2O_3 + 3C_2H_6 \tag{4.4}$$

In many respects, the use of ozone is similar to the use of an oxygen plasma, which produces a copious supply of highly reactive oxygen atoms.

When ozone decomposes it produces stable oxygen molecules (O_2), as well as oxygen atoms. In both cases, the oxygen atoms are highly reactive, and readily decompose the TMA.

The films were deposited at a temperature of 150–400°C. The growth rate of ozone process compared to water is slightly slower at deposition temperature of 300°C and above (0.6 - 0.85 Å/cycle). Below 300°C the growth rate is roughly the same and was observed to be even higher at 150°C [30]. As before, the films were amorphous, and very conformal. Ozone grown films were also observed to be rougher when deposited below 300°C [30]. The motivation for using ozone in this thermal ALD process was to minimize the amount of OH that might be found in the resulting films.

Specifically, when aluminum oxide is being considered as a replacement for the traditional thermal silicon dioxide gate oxide in an integrated circuit, its electrical behavior is very important. It appears [19], as shown in Figure 4.5, that the leakage current across a thin layer of Al_2O_3 deposited with ozone can be significantly less than when water is used.

However, the problem that remains with all attempts to replace thermal silicon dioxide on top of bare silicon, as a dielectric gate material, is that when the higher dielectric constant material is grown there is a tendency for some of the underlying silicon to oxidize during the process. For example [19], an interfacial layer of silicon dioxide 13Å thick is formed when growing a 50Å thick layer of Al_2O_3 using ozone.

Since the goal is to replace silicon dioxide layers as thin as 10–15Å, this interfacial layer is too thick. Although the interfacial layer appears to be much thinner when water is used [19], the leakage current (see Figure 4.4) is much higher.

4.2.1.3 *Processes with Other Precursors*

In addition to the use of TMA, $AlCl_3$ and DMACl as aluminum precursors, other studies have explored the use of other organometallic aluminum precursors that also contain oxygen [11, 20]. Among these are aluminum ethoxide $Al(OC_2H_5)_3$, aluminum propoxide $Al(OCH(CH_3)_2)_3$ and aluminum $(mmp)_3$ $Al(OC(CH_3)_2CH_2O(CH_3))_3$. In these studies the oxidizer used was either water [11, 20] or various alcohols [11].

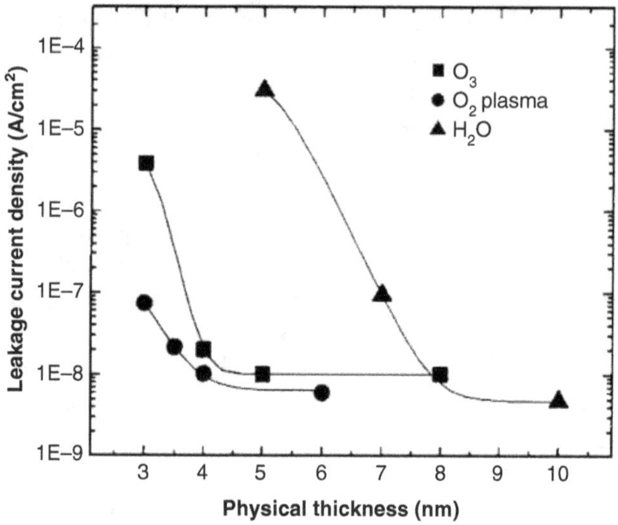

Figure 4.4 Leakage current density vs. Al_2O_3 film thickness deposited from TMA with different oxidants [19]. *Reprinted from Thin Solid Films, 476, S. C. Ha, E. Choi, S. H. Kim, and J. S. Roh, Influence of oxidant source on the property of atomic layer deposited Al_2O_3 on hydrogen-terminated Si substrate ,252, © 2005, with permission from Elsevier.*

From these studies it is not apparent that there was any specific benefit to be derived from replacing TMA with the particular precursors investigated. For example [20], aluminum oxide films grown from aluminum $(mmp)_3$ and water were inferior to those grown from TMA and water as far as dielectric properties were concerned. This could have been due to the carbon content, which was found to be 2.4%.

However, as noted in the previous section, there is a need to suppress the formation of an interfacial layer of silicon oxide when growing a thin layer of high dielectric constant oxide. Apparently, the use of isopropyl alcohol IPA $((CH_3)_2CHOH)$ as an oxidizer with TMA enables the growth of good quality aluminum oxide, with negligible carbon, and little interfacial oxide [21]. For example, when 129Å of aluminum oxide is grown at 250°C from TMA and water on bare silicon a 12Å thick interfacial layer of $SiAl_xO_y$ is formed. By contrast, when a 113Å film is grown from TMA and IPA at the same temperature there is no interfacial layer observed. Apparently, the IPA does not decompose on the original bare silicon surface at 250°C, whereas the water appears to. Unfortunately, after annealing at 800°C in argon for 5 minutes a

13Å interfacial layer was observed. Apparently, oxygen from the aluminum oxide layer was able to combine with the underlying silicon during the anneal.

Finally, studies have also been carried out when both reactants contain aluminum [31, 32]. For example, when oxide was grown from $AlCl_3$ and $Al(O^iPr)_3$, similar results were found, except that there was about 1.8% chlorine in the films.

4.2.1.4 *Radical Enhanced Processes*

In order to explore the Al_2O_3 films that can be deposited using an oxygen plasma, experiments were carried out at room temperature, with TMA as the source of aluminum. If good quality films could be grown this way, this would validate this approach. Of course, REALD oxides could also be grown at higher temperatures, if desired, to modify film characteristics. A simplified description of the chemical reaction is:

$$2Al(CH_3)_3\,(g) + 24O \rightarrow Al_2O_3\,(s) + 9H_2O + 6CO_2 \qquad (4.5)$$

Regardless of the details, clearly a great deal of atomic oxygen is needed to produce clean Al_2O_3 films.

Experiments were carried out in a downstream REALD system with an RF power supply driving a solenoid coil placed around a quartz tube. Again, the arrangement was similar to that described in Figure 4.7. Also, the RF frequency was 13.56 MHz.

In order to explore the quality of films deposited this way, 150–200Å thick films were grown, and they appeared to be smooth visually. To obtain more definitive information on the morphology of such films, a film was grown on silicon, and examined by TEM. In Figure 4.5, we see a 200Å thick smooth Al_2O_3 film on top of a layer of 80Å of SiO_2. Since the native oxide on the silicon wafer was probably no thicker than 20Å, the large amount of atomic oxygen available promoted oxide growth by diffusion through the SiO_2 layer. As this layer became thicker, the oxidation of the underlying silicon stopped. The Al_2O_3 film appears to be largely amorphous, with some unidentified crystalline structure.

The growth rate of this film was 1.8Å/cycle, which is ~ 2x the deposition rate of films grown by the thermal process. One explanation for this unusual rate of deposition is the possibility that the TMA film being adsorbed is composed of dimers, not monomers.

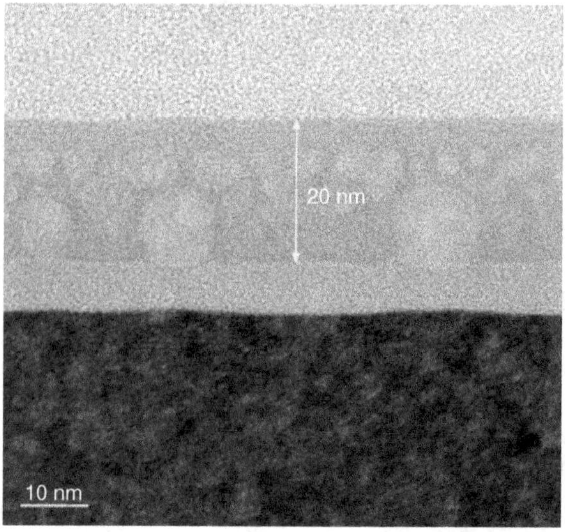

Figure 4.5 TEM of aluminum oxide film deposited at room temperature by REALD.

In fact, it has been known for many years that TMA vapor at room temperature is essentially all dimers [33].

Auger analysis confirmed that the films were indeed stochiometric aluminum oxide, with ~ 8 at.% carbon and 3 at.% nitrogen impurities.

The refractive index was determined to be 1.64, the dielectric constant 7.3, and the breakdown strength 7.5 MV/cm. These values indicate a surprisingly good quality aluminum oxide compared to both thermal ALD films and bulk oxide.

Additional studies have also been done, where this same process has been carried out in a capacitively coupled REALD reactor [34–37]. The RF power was in the range of 200–400 watts, and the reactor electrodes were ~ 30 cm in diameter. Also, presumably, the RF frequency was 13.56 MHz, although none of the references specified this value. As is well known from Plasma Enhanced CVD [38], the frequency at which a capacitively coupled PECVD reactor operates plays an important role in the quality of the film being formed. The same effect would be expected in REALD. Deposition temperatures ranged from 80–300°C. Of particular interest is the fact that in addition to the use of oxygen to create the oxidizing plasma, some nitrogen gas was added intentionally to see its effect on the film quality. The flows were O_2 (45 sccm) and N_2 (5 sccm),

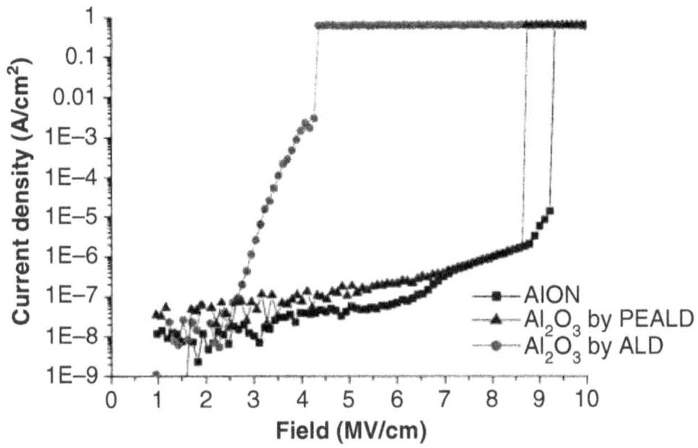

Figure 4.6 Current vs electric field for films deposited at 300°C [34]. © *2003 The Japan Society of Applied Physics.*

and the film grown would be better characterized as a doped aluminum oxide or Al_2O_3:N. In general these films contained less than 1 % of nitrogen.

In some applications, such as for the deposition of insulating layers in electroluminescent devices [34], these layers require high breakdown strength which must be maintained after a high temperature anneal. It turns out that Al_2O_3:N deposited by REALD exhibits a breakdown strength of over 9 MV/cm, which is better than that achieved with thermal ALD films. This behavior is illustrated in Figure 4.6.

First, we observe that the Al_2O_3:N films breakdown strength exceeds that for the room temperature film described earlier. It is possible that the carbon contamination in the room temperature film contributes to the reduction from 9 to 7.5 MV/cm. Secondly, both nitrogen doped films fare better than the thermal film at 3–4 MV/cm, although the 300°C REALD film with no nitrogen does almost as well.

The other issue is the stability of these films when annealed at a high temperature. In this case, the nitrogen doped film exhibits excellent thermal stability, compared to the nitrogen free material, as can be seen in Figure 4.7.

Because of the unique ability of the REALD technique to deposit these good quality films at, or close to room temperature, they may have application as passivation for organic devices. One example would be the passivation of organic light emitting diodes (OLED) [36]. Films grown at 60°C were shown to be quite effective in this

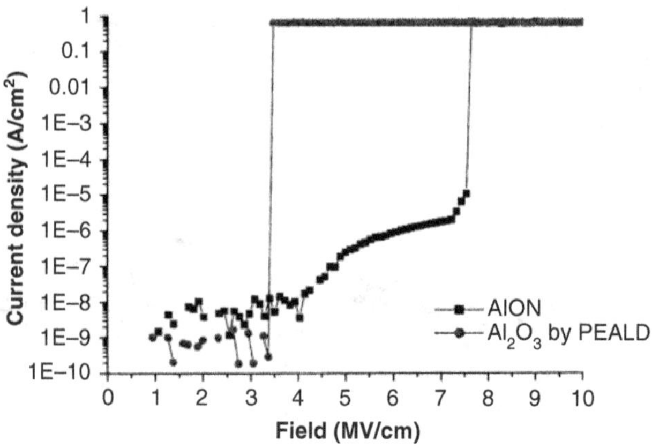

Figure 4.7 Breakdown strength for REALD Al_2O_3 and Al_2O_3:N films when annealed at 600°C for 5 minutes [34]. © *2003 The Japan Society of Applied Physics.*

application. In addition, there may be a need to deposit such low temperature dielectrics on plastic substrates for flexible display panels and plastic electronic devices [37]. In this reference, it was demonstrated that deposition of Al_2O_3:N onto polyethersulfone could be successful as long as the deposition temperature was no higher than 100°C. The major problem found was the distortion of the plastic substrate at higher temperatures than this. Clearly, the thermal expansion coefficients of a plastic substrate and an inorganic oxide layer will be quite different. When the oxide layer is deposited at a moderately high temperature, and then the coated substrate is cooled, it would be expected that the substrate would be subjected to a distorting stress. This problem can be avoided if deposition can be done at room temperature.

Since ALD films have shown potential for use as passivation and barrier layers, e.g. in OLED structures, it has inspired researchers to develop further the processes suitable for polymer substrates. Here the advantages of PEALD such as reduced substrate temperature and improved material properties are beneficial. PEALD aluminum oxide has been investigated by its water vapor and oxygen barrier properties on PEN [39, 40] and OLED passivation characteristics on PVK [41] substrates. In a study by Langereis *et al.* the deposition was carried out at temperature range of 25 to 200°C by using a remote oxygen plasma configuration in homebuilt ALD tool. The study revealed that it is possible to reduce significantly the ALD cycle time, compared to a thermal process, by using plasma, and at the same time reach reasonably good film performance as a water vapor diffusion barrier on PEN polymer substrate. Furthermore the study showed the relevance of deposition at room temperature in the case of a polymer substrate. When the deposition temperature was increased to 100°C, the barrier effect of the film started to deteriorate. This was attributed both to higher hydroxyl content in low temperature grown film prohibiting water permeation, and also to the mismatch of the thermal expansion of Al_2O_3 and the PEN substrate [39]. A Following study by the same group reported on film composition and growth characteristics. The film growth rate at room temperature was 1.8 Å/cycle and decreased with increasing deposition temperature. The carbon and hydrogen impurities in the film were at the level of 5 at.% and 15 at.% respectively, which both decreasing with increasing temperature. Interestingly the carbon content and the ratio of oxygen to aluminum in the film were decreased by doubling the oxygen plasma pulse length from 2 seconds to 4 seconds indicating the possibility of controlling the film stoichiometry by plasma assistance [40].

An approach to avoiding polymer substrate damage due to a high plasma power has been presented in one study, where electron cyclotron resonance (ECR) plasma was used in PEALD process. The film growth rate at room temperature was over 2 Å/cycle and depended on plasma power used. With plasma power of 100 W the growth rate around 1.6 Å/cycle was achieved. When plasma power was increased to 300 W the growth rate saturated to the highest level. Carbon content was less than 2 at. % with 100 W and decreased to 1 at. % with plasma power of 700 W. An 85 nm thick Al_2O_3 film was then investigated as the passivation layer in OLED device. It was shown that the lifetime of OLED device can be clearly increased by using Al_2O_3 ALD passivation layer [41].

4.3 Titanium Dioxide

Titanium dioxide is a widely investigated semiconductor mostly due to its photocatalytical properties. The surface reactions and photocatalytic mechanisms have been under focus since the discovery of photocatalytic splitting of water [42–50]. Its characteristics as a superhydrophilic coating was discovered in 1995 [49] which increased the possibilities of applying TiO_2 even further. As pigment TiO_2 has been used universally due to its high refractive index, color, and opacity in such products as paints, coatings, pastes, cosmetics, food, and medicine. Due to its hydrophilicity and strong oxidizing potential it has been applied especially in Japan for environmental purification as self-cleaning, and antibacterial surfaces, and indoor odor removal in construction cements, paints, tiles, and windows [50]. TiO_2 also has characteristics like high dielectric constant, good chemical and physical stability, and good transmission in the visible range which is why it can be used in multiple optical and sensing applications. Consequently TiO_2 has been extensively studied in waste water and air purification, in dye sensitized solar cells, gas sensors, and microelectronics.

An abundant number of studies have been conducted on photoactivity and photocatalytic properties of TiO_2. The two most studied crystal phases have been anatase and rutile and the characteristics of TiO_2 may be modified significantly by changing TiO_2 crystallinity, crystal phase and orientation. In the following a variety of precursors and ALD processes for TiO_2 have been presented. The emphasis has been on the crystalline structure and other characteristics of the films, especially with regard to applications.

4.3.1 Processes and Properties of TiO_2

4.3.1.1 The Halide Precursors and Processes

Titanium tetrachloride and water has been a popular ALD TiO_2 – process for a couple of decades. $TiCl_4$ is a liquid and can be evaporated at room temperature. It has been an easy choice for TiO_2 precursor since it is fairly volatile, stable, and reactive. It is also inexpensive which makes it a very good option for applications where the quality requirements allow some residual chloride.

For the tetrachloride-process the basic reaction mechanism is the following

$$TiCl_4 + 2\,H_2O \rightarrow TiO_2 + 4\,HCl$$

Various process temperatures have been studied for this halide process. In one of the first studies on ALD TiO_2 process, Ritala *et al* [51] used a temperature range of 150–600°C. The growth rate varied between 0.35–0.45 Å/cycle as a function of temperature when using Corning 1733 glass as a substrate. When depositing on soda lime glass the growth rate was 0.52 Å/cycle at 500°C. The chlorine residue in the film was also analyzed by RBS and 2 atomic-% of chlorine was found in the films deposited at 150°C. In the films deposited at 500°C the chlorine residue was below the RBS detection limit.

In the following years Aarik *et al* [52] observed that the growth of TiO_2 crystalline phases of anatase and rutile depended on the substrate temperature. Polycrystalline TiO_2 was grown on soda lime glass, fused silica, and (111)-oriented silicon when reaction temperature of ≥ 165°C was used. Even at 140°C hints of anatase growth had been discovered. At lower temperatures all films were amorphous. It was further discovered that the crystallinity was also thickness dependent so that the mimimun thickness where the crystals started to appear decreased when the reaction temperature was increased. At 210°C at the thickness of 10 nm the films were still amorphous but transformed to polycrystalline anatase at thicknesses of 15 nm–55 nm. At 300°C the film started growing as crystalline already at the thickness of 2 nm. At 350°C rutile phase started to appear, and at higher temperature the films were a mixture of anatase and rutile being rutile-dominating. In a later study [53] epitaxial films were also deposited with $TiCl_4$ and H_2O –process using α-Al_2O_3 (sapphire) as substrate material. At substrate temperature of 425°C the <101> plane of rutile was grown parallel to the <1$\bar{1}$02> plane of α-Al_2O_3. It was also noticed that anatase to rutile transformation took place at 350°C when using α-Al_2O_3 but needed temperature of 500°C when using SiO_2 as substrate. The epitaxial growth had an effect also on the refraction index; at 580 nm the index was 2.4–2.7 when the thin film was non-epitaxial but around 2.8 when the film was epitaxial.

The effect of substrate surface treatment to growth was studied by Finnie *et al* [54] and Mitchell *et al* [55]. They discovered that on hydrophilic RCA-treated Si <100> substrates TiO_2 grew initially amorphous transforming to crystalline at a critical thickness. Hydrophobic HF-treated Si <100> substrates produced island growth with small crystalline grains.

The chemical reactions of $TiCl_4$ – water process was studied by using a quadrupole mass spectrometer (QMS) to detect gas species evolving from the surface during film growth and a quartz crystal microbalance (QCM) to directly measure mass [56]. D_2O was used instead of H_2O in order to distinguish reaction byproducts in QMS. $TiCl_4$ and D_2O doses were varied at a temperature range of 150–400°C. It was noticed that DCl was generated both after $TiCl_4$ pulse and after D_2O pulse. The number of -OD groups which react with the entering $TiCl_4$ decreased with increasing temperature. Thus more DCl was released during at lower reaction temperature. At higher temperatures $TiCl_4$ molecule appeared to adsorb and then react with the D_2O pulse. At lower temperatures $TiCl_4$ seemed to decompose at the surface by losing one or more chlorine atom and then reacted with D_2O pulse releasing the remaining chlorine. The investigators concluded that beside the -OD/-Cl exchange, other reactions may take place and complicate the TiO_2 growth.

Photoactivity and photocatalytic properties were investigated by Kääriäinen *et al.* [48, 57]. Anatase – rutile transformation as a function of film thickness was scrutinized [48] at a reaction temperature of 350°C with soda lime glass as a substrate. At a thickness of 2 nm AFM-studies showed that crystalline grains were covering most of the surface. The AFM studies performed on the series of TiO_2 films with various thicknesses are presented in Figure 4.8. XRD spectrum showed that at 15 nm the film was anatase but the AFM showed that there were also facetted grains present that were most probably the beginning of rutile growth. The film transformed to rutile dominating between 65–130 nm. Photocatalytic activity was discovered to be the highest with anatase dominating films that were only 15 nm thick but contained a small amount of rutile. The investigators further observed that 10–15 nm TiO_2 that was grown at 250°C grew as polycrystalline rutile [46]. At 25 nm the rutile phase

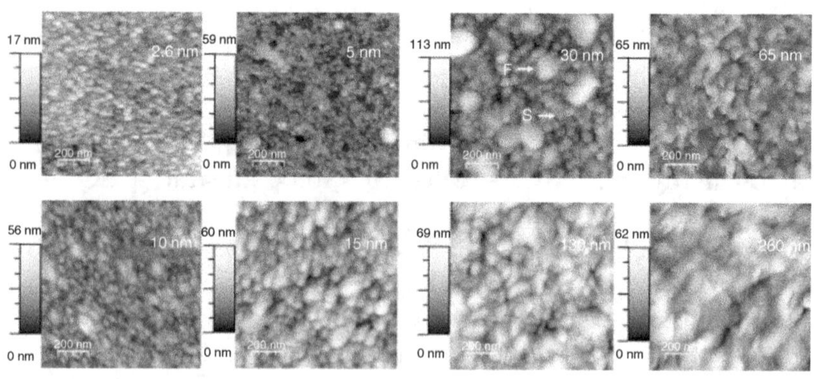

Figure 4.8 AFM images from TiO_2 deposited at 350 °C on soda lime glass. The film thicknesses are marked on the images [48]. *Reprinted from Thin Solid Films, 517, M.-L. Kääriäinen, T.O.Kääriäinen, D.C. Cameron, Titanium dioxide thin films, their structure and its effect on their photoactivity and photocatalytic properties, 6666, © 2009, with permission from Elsevier.*

has disappeared and a tiny XRD peak of anatase <101> was shown. At a thickness of 110 nm the film was polycrystalline anatase. Anatase or anatase dominant films were superior in photoactivity compared to rutile phase. In addition the investigators discovered that the anatase films were p-type and the rutile films were n-type. It was suggested that a good photocatalytic material would require an anatase-rutile p-n –junction to be able to perform well.

TiO_2 requires high photon energies for photoactivation due to the high bandgaps of 3.0 eV (rutile) and 3.2 eV (anatase). Nitrogen doping has been a successful way to lower the bandgap to visible light region. Pore *et al* [58] used ammonium gas (NH_3) in $TiCl_4$ – H_2O –process and alternated $TiCl_4$-purge-NH_3-purge and $TiCl_4$-purge-H_2O-purge in different proportions. The films were deposited onto borosilicate glass and indium tin oxide (ITO)-coated glass at 500°C. The films grew as polycrystalline anatase or anatase-rutile-mixture. In the study the bandgap was decreased from 3.3 eV to 2.9 eV but the doping weakened the photoactivity and the photocatalytic properties.

Also Kääriäinen and Cameron [59] applied nitrogen doping in the $TiCl_4$ –process by using ammonium hydroxide (NH_4OH) as oxidizer instead of water. The films with various thicknesses were deposited on soda lime glass at 350°C and the growth rate and crystal structure were similar to those in [48] The results from X-ray photoemission spectroscopy (XPS) showed evidence for interstitial nitrogen in the TiO_2 structure and the bandgap showed a minor red shift for the thickest samples. Film resistivity was also measured and the doped TiO_2 films possessed a higher conductivity than the undoped films which could be the result of nitrogen doping. Further the doping improved the photocatalytic activity by 2.5 times comparing to the best result with the undoped films. Likewise in the undoped films, the best photocatalytic result was found with 15 nm thick anatase dominant film which had a small amount of rutile. Photocatalytic activity of Pilkington Activ™ glass was also measured and the 15 nm thick nitrogen doped TiO_2 film performed 1.7 times better than Activ™. In addition the researchers discovered that similar to the undoped TiO_2 films [59], the anatase and anatase-dominating films were p-type and the rutile and rutile-dominating films were n-type.

One benefit stated for plasma or radical enhanced ALD is to lower the deposition temperature. This approach has been studied with an oxygen plasma in a $TiCl_4$ based ALD processes [60, 61]. The TiO_2 film was grown at the temperature range of 110 to 200°C with

RF plasma source and with 200W plasma power. The film growth rate of 1.35 Å/cycle was found to be constant at this temperature range. This growth rate is around 2.7 times greater than thermal ALD grown film using $TiCl_4$. Interesting finding in this study was that the film grown at 135°C has anatase structure defined by XRD. Furthermore the structural transformation was found to be film thickness dependent. The film grown at 200°C with the thickness of less than 20 nm was amorphous. Again the film with thickness of 54 nm showed anatase phase in both XRD and FTIR.

Another halide process using titanium tetraiodide has also been studied. TiI_4 is a solid precursor that has an adequate vapor pressure for ALD-process at 110°C. Kukli *et al* [62] studied TiI_4 and H_2O_2 –process depositing TiO_2 on soda lime glass and polycrystalline silicon at 250–490°C. At 325°C TiO_2 growth varied from 0.48–0.64 Å/cycle when the TiI_4 and H_2O_2 pulse varied from 0.2–0.5 s. The growth rate was dependent on the reaction temperature so that below 300°C the growth was linear. Between 300–400°C the growth slowed down and above 400°C the growth rate suddenly increased to 1.2 Å/cycle. In the XRD spectra the films grown on soda lime glass showed a tiny peak of anatase <101> below 300°C but the phase transformed to rutile <110> within 400–490°C. On polycrystalline silicon the film grew as anatase at 275°C, anatase-rutile mixture at 325°C, and polycrystalline rutile at 425°C. The iodine content was determined with time of flight elastic recoil detection analysis (TOF-ERDA) and the films did not contain any iodine residues. This has been commonly observed in all TiI_4-processes.

In their following study Kukli *et al* [63] used a QCM method to investigate the chemisorption process of TiI_4 and H_2O-H_2O_2 –process. It was observed that the TiI_4 adsorption was not totally saturative but TiI_4 was partially decomposed. The decomposition of TiI_4 increased between the reaction temperature of 200–300°C. The film formation mechanism was not noticeably affected by the pulse lengths or the growth cycles but depended mostly on the reaction temperature. Further study on TiI_4 and water was conducted in the temperature range of 135–375°C using Si<100> and SiO_2 as substrates [64]. The film was grown as anatase at all temperatures. In addition the investigators deposited TiO_2 on α-Al_2O_3 <012> achieving an epitaxial rutile phase at 445°C.

TiI_4 and O_2 –process was investigated at a temperature range of 230–460°C [65]. TiO_2 films were deposited on Si<100> and QCM was used to monitor the process. Again TiI_4 was partially decomposed into TiI_x. It was noticed that O_2 was able to react with TiI_x when the temperature was above 235°C and the oxygen pulse was 2 s. The deposited film was anatase <101> in all temperatures except at 460°C, where the film was rutile-dominating anatase-rutile -mixture.

Another halide precursor TiF_4 was studied by Pore *et al.* [66] Water was used as an oxidizer, and soda lime and borosilicate glass were used as substrate materials. Reaction temperature range was between 300–500°C. Saturated growth was reached at 400°C on both substrates. On soda lime glass the growth rates at 400 and 500°C were 0.8 and 1.0 Å/cycle, respectively, and for borosilicate 1.1 and 1.5 Å/cycle, respectively. Otherwise the growth varied greatly between the glass types and it was suggested that additional reactions occurred between TiF_4 and sodium species of the soda lime glass affecting the crystal structure. TiO_2 grown on soda lime at 400°C was anatase-rutile mixture and at 500°C it was polycrystalline rutile. On borosilicate all the films were polycrystalline anatase. In addition the films on soda lime were more uniform. However, at 400°C the growth on soda lime seemed to suffer from surface etching since the film thickness increased from the leading edge towards the end of the substrate. This was probably due to the corrosive nature of both TiF_4 and the by-product HF which probably had etched the film. EDX analysis was used to determine the residual fluorine although it was too small to be evaluated quantitatively. The estimation was 0–1 at-% for films deposited 400 - 500°C.

TiO_2 has also been found to be antibacterial. Narayan *et al* [67] coated nanoporous alumina with TiO_2. This type of nanostructure can be used for instance in drug delivery where it can be applied in controlled release of pharmacologic agents. Biocompatible and corrosion resistive TiO_2 was used as a coating since there is still debate whether aluminum is biocompatible. TiO_2 films of 4 nm and 9 nm were deposited at 300°C from $TiCl_4$ and water onto 20 nm and 100 nm pore size alumina substrates, respectively. XRD measurements showed the coatings were anatase and the film thicknesses were about 4–9 nm. It was found that TiO_2 coatings were as viable for human epidermal keratinocytes (HEK) growth as the plain porous alumina. In bactericidal tests conducted for *Staphylococcus aureus* and *Escherichia coli* the coated 20 nm pore size alumina inhibited the growth of both bacteria better than the coated 100 nm pore size alumina. The investigators suspected the difference in bactericidal tests was due to different photocatalytic activity or membrane-cell interaction with membrane surface.

4.3.1.2 *The Alkoxide Precursors and Processes*

Titanium alkoxide precursors are often preferred when halide residues are a problem either in the film or as a by-product. The first alkoxide studied in TiO_2 process was isopropoxide, $Ti(OCH(CH_3)_2)_4$ [68]. Isopropoxide has a relatively high vapor pressure and may be evaporated in an ALD system at 40 - 50°C. Chemically isopropoxide is stable up to about 325°C. In a study for $Ti(OCH(CH_3)_2)_4$ and H_2O –process, the temperature-independent growth of 0.3 Å/cycle was found between 250–325°C [68]. In the same study XRD spectra showed that the films grew as polycrystalline anatase peak <101> being the dominating one. The hydrogen residue was 0.3 at-% determined by nuclear reaction analysis (NRA). The thickness uniformity was found to be worse compared to films deposited with $TiCl_4$. Rahtu and Ritala [69] investigated reaction mechanisms for isopropoxide and deuterated water with a QCM and QMS -system. It was found that between 150–250°C and during the $Ti(OCH(CH_3)_2)_4$ pulse half of the isopropoxide ligands were released in the reactions with hydroxyl groups and another half was released during the water pulse. This is different compared to $TiCl_4$ –process where most of the ligands are released during the water pulse. The main peak for a by-product was for isopropanol, $(CH_3)_2CHOH$. At higher temperatures $Ti(OCH(CH_3)_2)_4$ began to decompose and for this reason 250°C was concluded to be a suitable process temperature.

Titanium isopropoxide was also deposited with H_2O_2 as an oxidizer [70]. Based on the QCM studies the adsorption of $Ti(OCH(CH_3)_2)_4$ was found to be a self-limited process at reaction temperatures of 100–250°C. At 100°C an unexpectedly high growth rate of 1.2 Å/cycle was discovered. This type of fast CVD-like growth of alkyloxides at temperatures near to precursor condensation temperatures has been lately found to be due to oligomeration of the alkyloxide groups.

When titanium isopropoxide and ozone were studied, the ALD reactor was equipped with an in situ attenuated total internal reflection Fourier transform infrared (ATR-FTIR) spectroscope [71]. It was found that after $Ti(OCH(CH_3)_2)_4$ pulse isopropoxy-ligand-terminated surface reacted with O_3 forming various surface carbonates releasing CO_2. The growth rate was found to be 0.5 Å/cycle being independent on the reaction temperature between 150 and 250°C.

Radical enhanced atomic layer deposition (REALD) was used to deposit TiO_2 from titanium isopropoxide and oxygen radicals [72] TiO_2 was deposited at 50°C on polymers and fibers, and at temperatures between 50 and 300°C on silicon, glass, platinum, and RuO_2 surfaces. The quality of the films grown at 50°C was not as good as in the films achieved with the thermal process. For the films at 50°C and 250°C the densities were 3.2 g/cm^2 and 3.8 g/cm^2, the refractive indices 2.2 and 2.4, and the residual carbon content 4 at.% and 0.4 at-%, respectively. All these films grew as amorphous.

TiO_2 from titanium ethoxide, $Ti(OCH_2CH_3)_4$ and H_2O was deposited on soda lime glass in a temperature range of 200–400°C [73]. $Ti(OCH_2CH_3)_4$ was evaporated at 80°C. The growth rate was not found to be self-limiting at any temperature. Between 250–300°C the growth rate varied from 0.25 to 0.44 Å/cycle. At 350°C titanium ethoxide was clearly decomposing causing a sudden increase in the growth rate. XRD analyses showed that within the range of 250–375°C all the films were polycrystalline anatase. In another study [74] on $Ti(OCH_2CH_3)_4$ and H_2O, TiO_2 was deposited on Si <100> and the growth was found self-limiting. At reaction temperature of 200°C TiO_2 was still amorphous after 1400 cycles but at 2000 cycles the film was growing as a very rough polycrystalline anatase with two major orientations of <101> and <004>. At 350°C the films were also polycrystalline anatase but much smoother with orientation <101> being the only dominating peak in XRD.

The first study for titanium tetramethoxide, $Ti(OCH_3)_4$ has been conducted by Pore *et al* [75] who also conducted the first photocatalytic measurements with ALD TiO_2 films. TiO_2 was deposited on borosilicate glass from $Ti(OCH_3)_4$ and H_2O at reaction temperatures between 200–400°C. The growth rate at 250°C was approximately 0.48 Å/cycle, carbon residue was low at 0.2 at-%, and the refractive index was measured to be 2.52. Above 300°C the precursor seemed to start decomposing which was seen as an increasing growth rate. The films deposited at 200°C were amorphous but at 250–300°C the films were polycrystalline anatase the dominating orientation by XRD being <101> peak. The investigators suspected that precursor decomposition started to disturb the crystal growth since at 350°C the anatase peaks had decreased. The photocatalytic tests were performed for a stearic acid film and for methylene blue –solution. The best performances were discovered with polycrystalline anatase films that were deposited at 275 and 325°C.

In order to study conformity of ALD TiO_2 Jõgi et al [76] applied $Ti(OCH_2CH_3)_4$ and H_2O with $TiCl_4$ and H_2O processes for creating TiO2 films into trenches. The substrates were 200 mm Si <100> wafers patterned with 7.5 μm deep elliptical holes etched in silicon with and aspect ratio of 1:40. Various reactor temperatures were used. The results are shown in Figure 4.9. The step coverage (conformity) was found to be excellent in all cases. At higher temperature the films were anatase dominant and rutile was found in films deposited at 500°C. The ethoxide films were found smooth but the thickness decreased towards the bottom of the trench.

In another study [77] TiO_2 was deposited with both $Ti(OCH_2CH_3)_4$ - H_2O and $Ti(OCH_3)_4$ - H_2O –process. The films were grown on Si- and Ti-substrates and the reaction temperature was kept between 250–300°C to avoid the decomposition of the titanium precursors. It was noticed that the films were amorphous during the first 400 cycles. During the next 400–3600 cycles the growth of polycrystalline anatase was discovered for both substrates. When comparing the alkoxides the investigators observed that the methoxide process produced a higher degree of crystallity in the films than the ethoxide process.

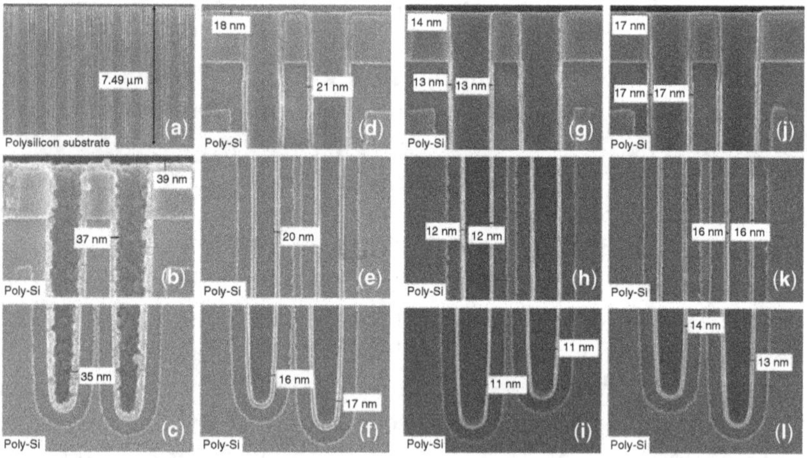

Figure 4.9 SEM images of (a) deep trenches used as substrates for conformal growth; (b), (c) films grown from TiCl4 at 500 °C at the top and in the bottom of trench, respectively; (d), (e) and (f) films grown from $TiCl_4$ at 125 °C, (g), (h) and (i) films grown from $Ti(OC_2H_5)_4$ at 300 °C, (j), (k) and (l) films grown from $Ti(OC_2H_5)_4$ at 175 °C at the top, middle and bottom of the trench, respectively. ALD cycle times were 2-2-2-5 s [76]. *Reprinted from Thin Solid Films, 516, I. Jõgi, M. Pärs, J. Aarik, A. Aidla, Matti Laan, J. Sundqvist, L. Oberbeck, J. Heitmann, K. Kukli, Conformity and structure of titanium oxide films grown by atomic layer deposition on silicon substrates, 4855, © 2008, with permission from Elsevier.*

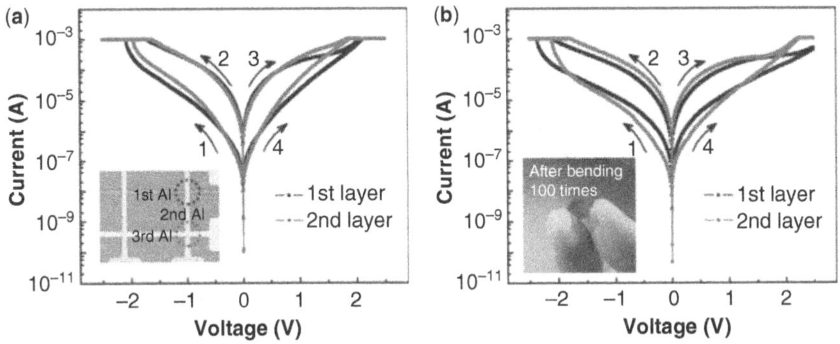

Figure 4.10 I-V curves of layered $Al/TiO_2/Al/TiO_2/Al$ structure on a PES substrate. Figure a represents the measurement without bending and Figure b the measurement after 100-times bending [78]. *Reprinted with permission from IOP-Publishing, © 2010*

The advantage of plasma-enhanced ALD has been used to study low temperature TiO_2 on PES polymer substrate to be used in future flexible resistive random access memory (RRAM) devices. The deposition was carried out at 80°C suitable to PES substrate. Amorphous TiO_2 film with 10 nm thickness carried out good mechanical flexibility demonstrated as a stable electrical property (I-V curves) after bending test shown in Figure 4.10 [78].

4.3.1.3 *Alkylamide Precursors and Processes*

Another option for replacing a halide process is to use alkylamide precursors. TiO_2 films have been deposited with tetrakis(dimethylamido) titanium, $Ti(N(CH_3)_2)_4$ (TDMAT) and various oxidizers such as H_2O, H_2O_2, H_2O-plasma, and O_2-plasma. Generally discovered characteristic of TDMAT processes has been a decreasing growth rate with increasing temperature. A proper explanation to this phenomenon is yet to be determined.

TDMAT and H_2O reactions can be written in two half reactions [79]

$$Ti(N(CH_3)_2)_4 + TiO_2\text{-}OH^* \rightarrow NH(CH_3)_2 + TiO_2\text{-}O\text{-}Ti(N(CH_3)_2)_3^* \quad (4.6)$$

$$TiO_2\text{-}O\text{-}Ti(N(CH_3)_2)_3^* + H_2O \rightarrow TiO_2\text{-}TiO_2\text{-}OH^* + 3(NH(CH_3)_2) \quad (4.7)$$

TDMAT processes have in general produced high purity TiO_2 films but several at-% of carbon impurities have been reported for O_2 plasma processes [80].

Titanium isopropoxide and TDMAT were compared as precursors using water as an oxidizer [80]. Isopropoxide was evaporated at 45°C and TDMAT was kept at 35°C for thermal stability. TiO_2 was deposited on Si<100> substrates. At 190°C the growth reached a plateau with isopropoxide pulsing time of 4 s while in TDMAT process the plateau was reached at 50°C with a pulsing time of 2s. All the TiO_2 films were found nitrogen and carbon free. For isopropoxide process the growth rate increased with increasing reaction temperature whereas with TDMAT the outcome was the opposite and the growth rate decreased within the temperature range of 35–330°C. The decomposition temperature for TDMAT is 180°C which can explain the lower slope down from 150°C. The growth rate decrease at low temperatures is still unclear. The investigators proposed an energetic model to explain the event and compare the different processes between the two precursors. The suggested model is not presented here.

Lee *et al* [81] investigated photoactivity and photocatalytic properties of TiO_2 films by applying plasma-enhanced ALD (PEALD) from a TDMAT and O_2 plasma –process. They deposited the films on Kapton (polyimide) sheet in a temperature range of 150–200°C. The growth rate was estimated to be 1.8 Å/cycle for all temperatures and all the films deposited under 200°C were amorphous. The films grown at 250°C were anatase. Photoactivity was tested under 10 W Black Light Blue lamp with a wavelength between 350–400 nm. Pilkington Activ™ glass was compared to the deposited films and both the anatase film and Activ™ glass possessed superhydrophilicity in 30 minutes of exposure time. In photocatalytic tests the degradation of 4-chlorophenol (4-CP) under UVwas followed for 5 hours. The amorphous films were not photocatalytic, the Pilkington Activ™ showed some activity, and the anatase film performed the best in the test.

Lee *et al* [82] further used the same process and deposited TiO_2 on galvanized iron substrates at a temperature range of 200–300°C. The investigators reported high growth rates of 2 Å/cycle. Films deposited at 200°C were amorphous and above that temperature anatase. The films were tested as photocatalysts by degrading 4 -CP under UV-light. The most crystalline anatase (300°C) film was found to be the most effective.

TiO_2 has been successfully deposited on polymers with temperature processes. Amorphous TiO_2 was deposited on poly(methylmethacrylate), PMMA at 60°C from TDMAT and ozone [83]. TiO_2 was shown to protect PMMA substrate from UV – light and improve adhesion between PMMA and sputtered titanium and titanium carbide coatings. In another study Kääriäinen *et al* [84] scrutinized the plasma characteristics of PA-ALD in order to minimize the detrimental plasma conditions for sensitive substrates such as photoresistive PMMA. In the depositions amorphous TiO_2 was grown on PMMA, polycarbonate (PC), and polypropylene (PP) substrates at temperatures between 50–70°C with the non-detrimental process conditions to polymer substrate.

4.3.1.4 *Heteroleptic Precursors and Processes*

TiO_2 has a high dielectric constant and is therefore potential material for insulators in the semiconductor industry in particular as dielectrics in complementary metal oxide semiconductor (CMOS) transistors, and in dynamic random access memory (DRAM). Especially TiO_2 rutile phase is desirable for its electrical properties.

Additionally an enhanced interest for ternary and mixed oxides has increased the demands for metal oxide precursors. Lately heteroleptic precursors for TiO_2 with high volatilities and reactivities, and thermal stability have gained interest. In addition a high purity for the deposited films may be achieved. Ozone or oxygen plasma processes have been applied in these processes since precursors like cyclopentadienyl need stronger oxidizers than water or hydrogen peroxide.

Sarkar *et al* [85] applied a heteroleptic precursor where an alkylamido ligand has been attached with cyclopentadienyl ligand in order to gain the benefits from both chemical groups. They studied methylcyclopentadienyl alkyamido titanium $[Ti(C_p^{Me})(NMe_2)_3]$ and oxygen plasma process at a temperature range of 100–300°C. The growth rate was found to be 0.7. - 0.8 Å/cycle which was higher than with other organometallic precursors. In addition there was no nucleation delay to be seen. The films grew as amorphous between 100–200°C and at 300°C the films were anatase. Analysis with Rutherford backscattering (RBS) showed that the elemental residues were reasonably low; for the films deposited at 200°C carbon, hydrogen, and nitrogen contents were 0.6 at-%, 2 at-%, and 2.3 at-%, respectively.

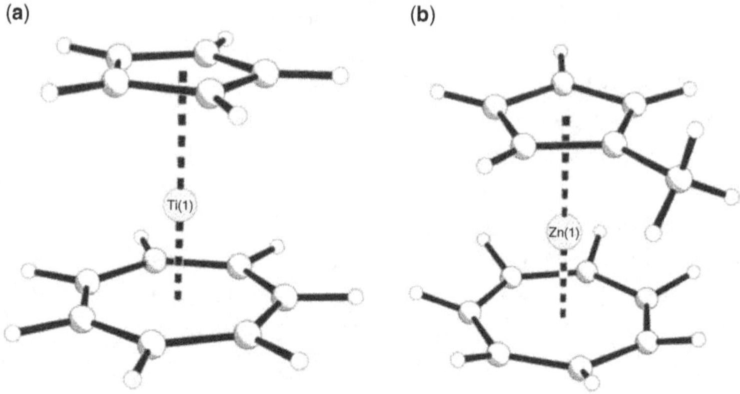

Figure 4.11 Schematic pictures of a) η7-cycloheptatrienyl-η5-cyclopentadienyltitanium (CpTiCHT) and b) η7-cycloheptatrienyl-η5 methylcyclopentadienylzirconium (CpMeZrCHT) [86]. *Reprinted with permission from J. Chem. Mat. 24, Niinistö, T. Hatanpää, M. Kariniemi, M. Mäntymäki, L. Costelle, K. Mizohata, K. Kukli, M. Ritala, M. Leskelä, 2002 © 2012 American Chemical Society.*

In another study [86] with heteroleptic TiO_2 precursors, cycloheptatrienyl (CHT, C_7H_7) ligand was attached with cyclopentadienyl (Cp, C_5H_5) ligand in synthesis for a novel titanium precursor of $(C_5H_5)Ti(C_7H_7)$ or CpTiCHT. Figure 4.11 presents heteroleptic precursors for TiO_2 and for ZrO_2 (will be discussed in Chapter 4.5.1).

In this study TiO_2 was deposited by using CpTiCHT and ozone at 250–450°C [86]. The growth rate increased throughout the range. The researchers stated that this type of increasing growth has been detected often in ozone processes. At 400°C the growth rate increased with increasing precursor pulse indicated partial decomposition of the precursor. The investigators suggested 350°C to be a suitable reaction temperature for CpTiCHT and O_3 –process. At this temperature the growth rate was 0.6 Å/cycle.

The films were deposited on Si substrate. Based on the XRD patterns the films deposited at 250C were polycrystalline anatase. At 300C the films were first anatase but at a thickness of 39 nm also rutile <110> appeared. At 325C the films became rutile dominating at a film thickness of 74 nm.

Usually in other precursor processes rutile is difficult to deposit especially on crystalline substrates like silicon and require a high reaction temperature but with CpTiCHT and O_3 the investigators were able to grow rutile at fairly low temperatures.

Due to its high dielectric constant TiO_2 films may be used as gate dielectrics in the field effect transistors. Jõgi *et al* [87] deposited TiO_2 from $TiCl_4$ and $Ti(OCH_2CH_3)_4$ with H_2O and H_2O_2 as oxidizers. The Mo–TiO_2–Al stacks were prepared on SiO_2 layer deposited on Si (1 0 0) substrate. The deposition temperature was between 175–225°C. From all the precursor variations $TiCl_4$ and H_2O –process produced the highest crystallinity in the films whereas the lowest was found with the $Ti(OCH_2CH_3)_4$ and H_2O_2 -system. In a consequent study [88] the researchers discovered that a highest permittivity and lowest equivalent oxide thickness was achieved with the $TiCl_4$ and H_2O –process while the lowest permittivity and the lowest leakage current values were carried out with the $Ti(OCH_2CH_3)_4$ and H_2O_2 –process. The Mo–TiO_2–Al stack is presented in Figure 4.12.

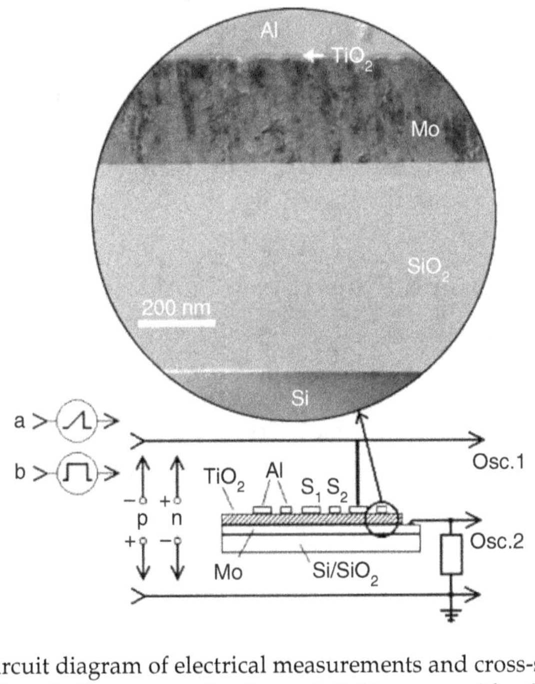

Figure 4.12 Circuit diagram of electrical measurements and cross-section of a Mo–TiO2 –Al stack. To measure the dynamic J –V curves with a 2-channel oscilloscope (Osc.1 and Osc.2), (a) rising voltage ramp and (b) constant voltage (p—positive polarity or n—negative polarity on the Mo electrode) were applied [88]. *Reprinted from Mat. Sci. Semicon. Proc. 9, I. Jõgi, K. Kukli, J. Aarik, A. Aidla, J. Lu, Precursor-dependent structural and electrical characteristics of atomic layer deposited films: Case study on titanium oxide, 1084 © 2006, with permission from Elsevier.*

The interest in ALD titanium dioxide studies will unquestionably continue due to its many possibilities in various applications.

4.4 Zinc Oxide

Zinc oxide (ZnO) is a semiconductor with a wide direct bandgap of circa 3.3 eV which is close to that of TiO_2. A hexagonal wurtzite structure is very common for ZnO. Due to its high light scattering ability with refractive index of 2.0, and absorbance in upper UVA-region wavelengths it can be applied in many UV-blocking systems. Its native doping form n-type has been suspected to arise from oxygen vacancies or interstitial Zn atoms. It possesses a good transparency, high electron mobility, and a strong room temperature luminescence. ZnO has many applications in optics and optoelectronics like transparent conducting electrodes in liquid crystal displays and solar cells, electronics in thin-film transistors, and light –emitting diodes as well as in energy saving or heat protecting windows. ZnO is also bactericidal and therefore can be used in many medical and antiseptic applications. Consequently the interest in modifying and developing ZnO thin films has grown during the recent years and this can be seen also in the increasing number of ALD studies.

4.4.1 Processes and Properties of ZnO

4.4.1.1 Diethyl Zinc Process

The most commonly used ZnO precursors in ALD has been diethyl zinc (DEZ) due to its high reactivity with water.

The binary CVD reaction for ZnO deposition is

$$Zn(CH_2CH_3)_2 + H_2O \rightarrow ZnO + 2C_2H_6$$

this can be written in two ALD half-reactions [89]

$$OH^* + Zn(CH_2CH_3)_2 \rightarrow OZn(CH_2CH_3)^* + C_2H_6 \quad (4.8)$$

$$OZn(CH_2CH_3)^*\ H_2O \rightarrow OZnOH^* + C_2H_6 \quad (4.9)$$

Lim and Lee [90] studied the process at temperature range of 100–250°C depositing films on sapphire <0001> substrates. Pulse lengths for DEZ and H_2O were 0.2 s and 0.15 s, respectively. Nitrogen (N_2) was used as a purging gas for 2 s after each pulse. An ALD window was found for reaction temperatures between 130–180°C where the growth rate was about 2.6 Å/cycle. The preferred crystalline orientation for the films was <002> which is also the thermodynamically favorable orientation for ZnO. In addition the crystallinity was found to improve with increasing temperature.

Kim *et al* [91] investigated a DEZ and ozone (O_3) process and compared it to the DEZ – H_2O process. The films were deposited on SiO_2/Si substrates at a temperature range of 230–300°C. At 250°C the saturation limit was reached with 1 s DEZ pulsing time and the growth rate was 1.7 Å/cycle which was lower than frequently reached value of 2 Å/cycle with DEZ-H_2O process. In both precursor processes the preferred crystal orientation was <002>. The orientation of <002> is particularly desired in surface acoustic wave (SAW) devices [92]. Kim *et al* also measured the resistivities of the films and found the values of O_3-deposited films to be 10–1000 times higher compared to those deposited with H_2O. The researchers expected this was due to Si diffusion from the substrate during the ozone process.

Schuisky *et al* [93] studied in situ resistivity of ZnO with a four-point-probe during the growth of the film. The films were deposited with DEZ and H_2O at 177°C (450 K) with pulsing times of 1 s and purging times of 5 s. The growth rate was found to be 2.0 Å/cycle. The resistivity increased considerably after every DEZ pulse and decreased after every H_2O pulse. The increase can be explained due to the conversion of absorbed hydroxyl groups into the alkyl species of the DEZ precursor. After each H_2O pulse a new ZnO layer is formed again decreasing the resistivity during the growth.

ALD grown ZnO was compared with sputtered ZnO by Lim *et al* [94]. DEZ and water were used at reaction temperatures between 100–250°C using Ar as purging and carrier gas. The films were grown on Corning 1737 glass and Si <001>. The growth rate increased with increasing temperature and decreased again at 175°C and higher temperatures. At 200°C the saturated growth rate was 1.7 Å/cycle and was achieved at a pulse time of 2 s on both substrate types. The films were polycrystalline at all temperatures with a dominant orientation of <002>. Resistivity and mobility measurements showed that at low growth temperature ($T \leq 125°C$)

ALD grown ZnO films had high resistivity of 10 Ω cm and low mobility of 3 $cm^2/V{\cdot}s$ whereas the films prepared at higher temperatures showed lower resistivity of 0.02 Ω cm and higher mobility of 15 $cm^2/V{\cdot}s$. In contrast sputtered ZnO films had much higher resistivity all above 1 kΩ cm and mobilities of 1.5–3.3 $cm^2/V{\cdot}s$. The researchers concluded that growth parameters change the electrical properties a great deal and that both ALD and sputtering are potential processes for producing TFTs.

Fong *et al* [95] studied ZnO growth in situ with synchrotron x-ray scattering and x-ray fluorescence (XRF) system. They used diethyl zinc (DEZ) and water as precursors and Si as a substrate. The investigators used exceptionally long pulsing and purging times to be able to follow the growth. DEZ (1 min) – purge (5 min) – H_2O (10 s) – purge (30 min). They discovered that ZnO grew first as islands and coalescence occurring at fourth growth cycle. In addition surface smoothing was discovered during the purges of the first few cycles. The growth rate for cycles between 6–10 was 4.2 Å/cycle and after 10 cycles it decreased to 3.0 Å/cycle.

Aluminum doped ZnO has been found to be a plausible option for transparent conducting oxides like tin-doped indium oxide (ITO) or fluorine-doped tin oxide (FTO). Aluminum doping of ZnO films was studied by presenting precursors sequentially into ALD reactor and varying the amount of cycles for each precursor [96]. ZnO:Al films were deposited at 125°C with DEZ, TMA, and H_2O as precursors and Ar was used as a purging gas. Figure 4.13 presents different variations in Al-doping of ZnO by sequencing the precursors.

The growth of ZnO:Al films was then scrutinized by using in situ quartz crystal microbalance and electrical conductance analysis. Infrared transmission analysis confirmed Zn-O-Al bonds and the amount of Al-bonds increased when TMA pulse was ended with H_2O pulse instead of DEZ pulse. The amount of Al-bonds decreased when DEZ and TMA pulses were introduced right after another. The investigators suspected that the zinc-organic species were removed from the surface after the TMA pulse.

King *et al* [97] deposited ZnO films with thicknesses up to 30 nm on 550 nm diameter SiO_2 spheres to be applied in UV blocking cosmetics. DEZ and water were used as precursors in a fluidized bed reactor at 177°C and the growth rate was found to be 2.0 Å/cycle. Figure 4.14 presents A STEM image of a coated SiO_2 particle along

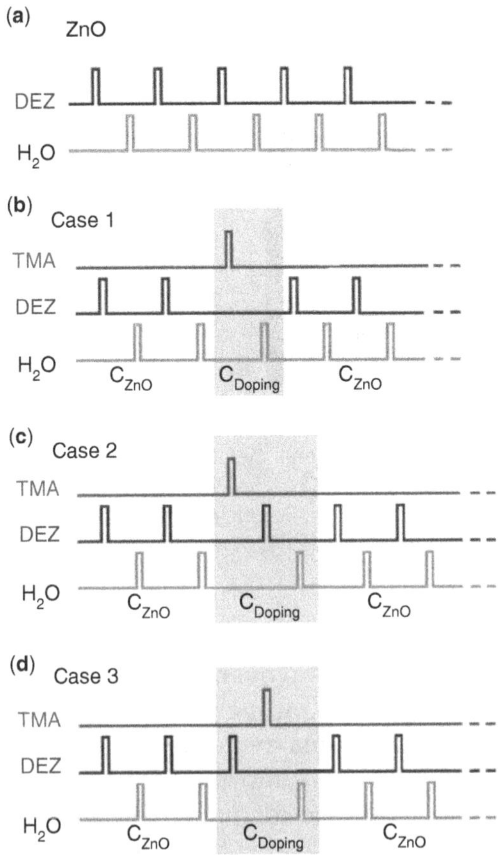

Figure 4.13 Schematic diagram of the process sequences used to grow (a) undoped and (b)-(d) Al- doped ZnO films. The precursor pulse and Ar purge time is fixed at 1 and 20 s, respectively. Three different doping sequences are classified as (b) case 1, (c) case 2, and (d) case 3 [96]. *Reprinted with permission from Chem. Mater. 21, J. S. Na, Q. Peng , G. Scarel , G. N. Parsons, 5585 © 2009 American Chemical Society.*

with UV absorbance of two different thicknesses of ZnO on SiO_2 particles.

In addition a few nanometers of ZnO were deposited on TiO_2 nanoparticle. The ZnO was found to be polycrystalline and further annealing of the ZnO – TiO_2 particulates at 600°C produced zinc titanate. The researchers found no particle agglomeration due to the functionalization of TiO_2.

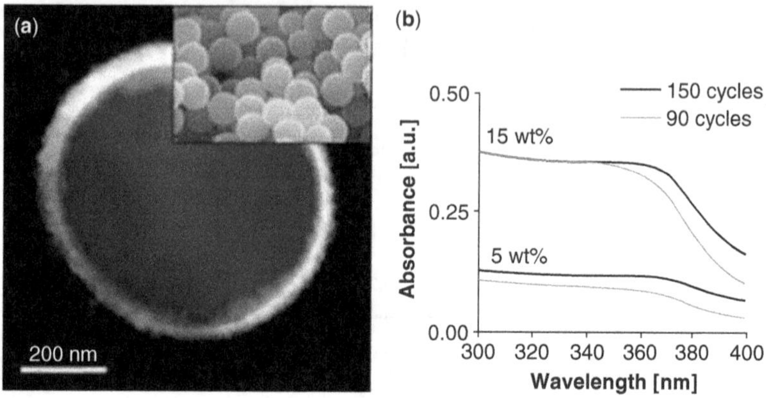

Figure 4.14 a) STEM image of 550 nm SiO_2 sphere coated with 150 cycles of ZnO. SEM image (inset) of uncoated spheres. b) UV absorbance of dispersed SiO_2 spheres coated with 90 and 150 ZnO cycles. The equivalent loading of uncoated SiO_2 is used as the baseline for each. Scattering is neglected using this method. *Reprinted with permission from Wiley-VCH Verlag GmbH & Co © 2008.*

Coatings on fibers such as woven fabrics or paper have gained plenty of interest recently. By adding photoactivity or photocatalytic properties, or electrical or magnetic functionality, a high surface area fiber mat can be turned into photochemical devices, batteries, sensors and capacitors. Cotton paper and fiber mat were coated by ZnO at 115°C using diethylzinc (DEZ) and water as precursors [98]. On planar silicon substrate the growth was linear being 2.1 Å/cycle. ZnO on both cotton and paper showed effective conductivity of up to 24 S cm^{-1}. It was concluded that ALD can be regarded as a feasible tool to attain well-controlled conductive and semiconductive coatings on various fiber materials.

Oxide semiconductors are a potential alternative to amorphous or polysilicon in thin film electronics. Mourey *et al* [99] deposited crystalline ZnO by plasma-enhanced atomic layer deposition (PEALD). DEZ and N_2O were used as precursors at 200°C. Nitrogen doping effect was desirable in order to gain lower resistivity ZnO film. The regular ALD cycle was altered so that N_2O was used to purge the DEZ excess and RF plasma was then used to oxidize the adsorbed DEZ finally purging the excess molecules by N_2O. ZnO was deposited on PEALD Al_2O_3 and the growth rate was found to be 2.5 Å/cycle. The investigators reported having fabricated high-mobility TFTs with electron mobility between 20–30 $cm^2/V{\cdot}s$.

Finally, another precursor methylzinc isopropoxide was used with water in a study where an attempt was made to decrease the conductivity of ZnO by doping the film with aluminum [100]. Dimethylaluminun isopropoxide was used as doping precursor. The growth rate of ZnO was found to be 1.9–2.0 Å/cycle with the temperature range of 160–200°C. Room temperature photoluminence studies were also conducted and the researcher found a free excitonic peak at 3.27 eV. The doping decreased the resistivity from 3.2–5 Ω-cm to about 1.4 Ω-cm.

4.5 Zirconium Dioxide

Zirconium dioxide is fairly inert chemically; it is attacked only by fluoride and alkalis. It is tough and abrasive, as well as corrosion resistant. It can withstand high temperatures up to 2400°C and it becomes electrically conductive above 600°C. Additionally its crystal structure also allows oxygen to more freely through it. Therefore it can be used in oxygen sensors and fuel cell membranes. ZrO_2 has a high dielectric constant and a high bandgap of 7.8 eV therefore it has potential to replace SiO_2 in metal-oxide semiconductor field effect transistors (MOSFETs) and to be used generally in complementary metal oxide semiconductors (CMOS).

In order to improve integrated circuit performance continued thinning of the gate oxide (SiO_2) has been attempted. However, once the gate oxide thickness drops to 15Å, even at a 1 volt operating voltage, leakage current can be as high as 1 amp/cm^2. Thinner gate oxide provides a higher gate capacitance. Therefore if a metal oxide layer with a much higher dielectric constant could be used, a thicker layer than 15 Å would be allowed and the leakage current would be less of a concern. In the following we look at several examples from the development of ALD ZrO_2 thin films.

4.5.1 Processes and Properties of ZrO_2

4.5.1.1 The Halide Precursors and Processes

In the first reported ALD results [101] ZrO_2 film was grown from $ZrCl_4$ and water at 500°C. The films were deposited on soda lime glass and nitrogen was used as a purging gas. The growth rate was found to be 0.53 Å/cycle and the films were amorphous containing

0.5% chlorine contamination, and less than 0.4% hydrogen. In other studies ZrO_2 was deposited at 300°C onto silicon wafers and the films were found polycrystalline [102].

In one later study [103] ZrO_2 was deposited onto glass substrates within a temperature range of 250 to 450°C and the ALD window was found over the broad range of 280–350°C. The growth rate was doubled compared to study in [101]. The films were found to be polycrystalline.

Aarik *et al* [104] and Kukli *et al* [105] applied a wider temperature range of 180–600°C. Water and hydrogen peroxide were used as oxidizer. The film structure was studied using reflection high-energy electron diffraction (RHEED). It was observed that, as shown in Figure 4.15, for each growth temperature there was a film thickness at which crystalline phases began to appear. At low growth temperatures amorphous films could be grown as thin as 100Å.

ZrO_2 was studied as a replacement for the SiO_2 used as the gate oxide in CMOS silicon integrated circuits. Films as thin as 40Å were

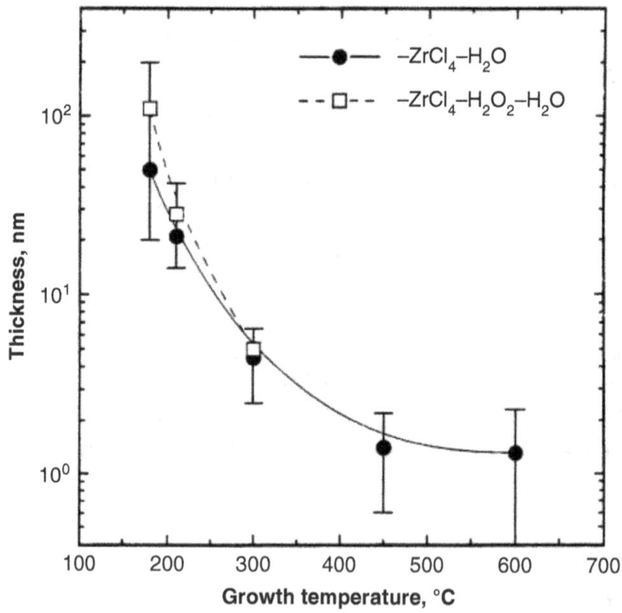

Figure 4.15 Film thickness at which crystalline phases appear as a function of deposition temperature. Films were deposited onto Si(100) substrates.[104] *Reprinted from Thin Solid Films, 408, J. Aarik, A. Aidla, H. Mandar, T. Uustare, and V. Sammelselg, Growth kinetics and structure formation of* ZrO_2 *thin films in chloride-based atomic layer deposition process, 97, © 2002, with permission from Elsevier.*

deposited from $ZrCl_4$ and H_2O at a 300°C [106]. ZrO_2 was deposited onto p-Si(100) substrate which had a 12 Å native oxide layer of SiO_2. The 40Å thick ZrO_2 film was deposited in 80 s with cycle times of 2 s. The structure analysis with high-resolution transmission electron microscopy (HRTEM) and composition analysis with PEELS (parallel electron energy loss spectroscopy) showed that the 40Å thick upper layer consisted of amorphous $(ZrO_2)_x(SiO_2)_{1-x}$ with x>>0.5, and some ZrO_2 nanocrystals. Since the layer was grown as amorphous, an annealing would be needed.

ZrI_4 and H_2O_2 process was studied within a temperature range of 250–500°C [107]. The films were deposited onto Si(100) from which the native oxide had been removed with HF etch. An ALD window was found in temperature range of 350 - 450°C and the growth rate was 0.6 Å/cycle. XPS measurements indicated that the iodine content of these films ranged from 1.3% at 250°C to 0.8% at 350°C. The thickness of these films was approximately 500–700 Å therefore they were too thick to be considered for application to CMOS processes. Thinner films were reported in a later study of the same author [108].

4.5.1.2 *The Alkoxide and Alkylamide Precursors and Processes*

All the following examples of ZrO_2 studies have had the same goal to achieve a nanometer thick ZrO_2 film with as high electric constant as possible to be suitable for CMOS applications. In the discussion of TiO_2 processes (Chapter 4.3.1) it has been already discovered that alkoxides have a limited thermal stability and the ALD window is hard to reach or it is too narrow. In the following alkoxide studies the same problem was recognized.

Kukli *et al* [109] studied the suitability of tetra-*tert*-butoxide Zr $[OC(CH_3)_3]_4$ as a new ZrO_2 precursor. Water was used as oxidizer over a temperature range of 150–300°C. The precursor apparently decomposed since the growth rate varied as the temperature was raised. Despite the fact that the films possessed a very high dielectric constant of 32, they contained substantial hydrogen and carbon contamination. In another study [110] Zr $[OC(CH_3)_3]_4$ was also found to be very moisture sensitive and was likely to decompose even during storage.

These thermal ALD studies of ZrO_2 indicated that deposition with Zr $[OC(CH_3)_3]_4$ and water, over the temperature range 150–300°C, did not demonstrate any temperature window. Subsequent thermal experiments with Zr $[OC(CH_3)_3]_4$ using oxygen [111, 112], indicated that at 250°C ALD films could also be grown, but in this case there was no information regarding deposition at other temperatures. Also, there is no information as to the deposition rate as a function of oxygen atom flux.

Two experiments [111, 112], attempted radical enhanced deposition of zirconium oxide from two different precursors (Zr $[OC(CH_3)_3]_4$ and $Zr(NEt_2)_4$). The oxygen atoms were generated in a downstream RF plasma reactor operated remotely at 13.56 MHz. The power level was set at 100 watts. Deposition was carried out on bare silicon wafers at one temperature, 250°C.

Deposition from Zr $[OC(CH_3)_3]_4$ showed a growth rate of 2.8 Å/cycle, and yielded films with 3–4 % carbon. From $Zr(NEt_2)_4$ the films were grown at a rate of 1.3 Å/cycle and had 1–3 % carbon. Both precursors yielded stoichiometric ZrO_2. Also, both films showed a smooth interfacial layer of 20–30 Å thickness when 40 Å of film was deposited. The grown ZrO_2 appeared amorphous, and had a smooth surface. Indications were that the interfacial layer was a zirconium silicate, which was thick enough to stop oxygen diffusion down to the silicon wafer. Finally, leakage current density was lowest for the films grown from $Zr(NEt_2)_4$, and was 2.5×10^{-9} A/cm ^{2} at –1.5 V.

Similar experiments were carried out to deposit both ZrO_2 and HfO_2 from Zr $[OC(CH_3)_3]_4$ and Hf $[OC(CH_3)_3]_4$, using an electron cyclotron resonance (ECR) reactor in a downstream configuration to generate oxygen atoms [113, 114]. In this case, they lowered the deposition temperature to 200°C, because they claimed that deposition at 250°C did not appear self-limited. In this study, the authors also show that HfO_2 films grown with an oxygen plasma contained less carbon than a similar film grown with oxygen but no plasma. As in the previous study, they also found that the leakage current through a ZrO_2 film is quite low. Presumably, this behavior can be attributed to the amorphous nature of the film, as well as the presumably 20 Å thick zirconium silicate under layer.

Zirconium aminoalkoxide precursors $Zr(dmae)_4$, $Zr(dmae)_2(OtBu)_2$, and $Zr(dmae)_2(OiPr)_2$ (dmae = dimethylaminoethoxide $[OCH_2CH_2N(CH_3)_2]$) were tested for ALD [115]. The investigators noticed the same problem as in earlier studies of alkoxides, that no ALD window was reached. Evidently,the precursors were decomposing during the deposition, and hence they contained high amounts of residual hydrogen and carbon.

Three compounds of alkylamides were tested as ZrO_2 precursors with water as an oxidizer. 116 They were tetrakis(dimethylamino) zirconium, $Zr(NMe_2)_4$; tetrakis(ethylamino)zirconium, $Zr(NMeEt)_4$; tetrakis(diethylamino)zirconium, $Zr(NEt_2)_4$. The precursors are liquid at room temperature and possess a sufficient vapor pressure for deposition. Depositions were reported to be done at 50 - 500°C reaction temperatures with deposition rate about 1 Å/cycle. The films deposited in a temperature range of 250–350°C contained significant amounts of carbon and nitrogen residues which was probably due to precursor decomposition. Otherwise the films were found to be conformal the refractive index being 2.1. Precursor decomposition of alkylamides has been also reported by Päiväsaari *et a*l [117].

4.5.1.3 *Cyclopentadienyl, Metallocene, and Heteroleptic Precursors and Processes*

Cyclopentadienyl and metallocene compounds are volatile and highly reactive and therefore well suited for ALD processes. Nevertheless they have not been applied often in high-*k* gate oxide depositions. Heteroleptic precursors were also mentioned in Section 4.3.1. They have high volatilities and reactivities, and they possess a good thermal stability. The films are expected to have a high purity.

Niinistö *et al* [118] examined the cyclopentadienyl process for ZrO_2 thin films with thicknesses below 20 nm. ZrO_2 was deposited onto HF-etched or native oxide containing Si(100) substrates at 350°C. An organometallic precursor, $Cp_2Zr(CH_3)_2$ (Cp = cyclopentadienyl, C_5H_5) was used as a precursor with water or ozone as oxygen source. The ozone process produced an interfacial layer with intermixing SiO_2 and ZrO_2 between the very uniform and dense ZrO_2 layer and the Si substrate. The water process was strongly affected by the substrate. The growth onto the HF-etched silicon substrate resulted in a growth of 2.9 nm interfacial layer with intermixing of SiO_x and ZrO_2 along with an islandtype growth of ZrO_2. The HRTEM images are presented in Figure 4.16.

A retarded growth rate was recognized for the initial growth in water process. This was due to the lack of reactive surface sites on the H-terminated Si. This further resulted in a higher leakage current compared to films from the ozone process. The films from the ozone process showed also low hysteresis and nearly ideal flatband voltage.

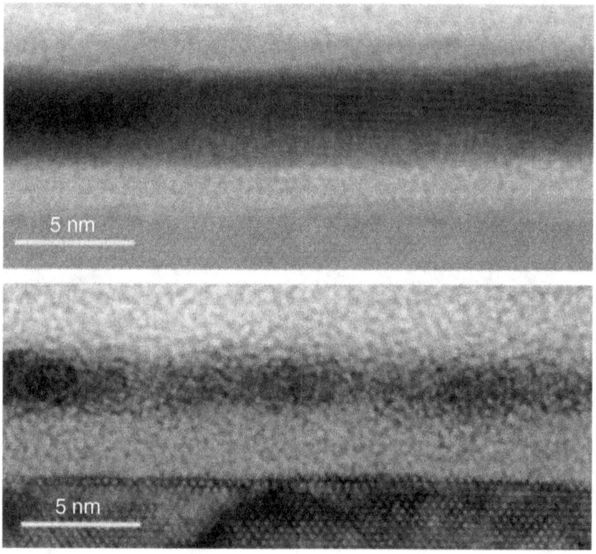

Figure 4.16 HRTEM images of the ZrO_2 films deposited onto HF-etched Si(100) by the (top) $Cp_2Zr(CH_3)_2/O_3$ (ZrO_2 thickness: 3.8 nm) and (bottom) $Cp_2Zr(CH_3)2/H_2O$ (ZrO_2 thickness: 2.1 nm) process [118]. *Reprinted with permission from J. Appl. Phys., 95, J. Niinistö, M. Putkonen, L. Niinistö, K. Kukli, M. Ritala, M. Leskelä, Structural and dielectric properties of thin ZrO_2 films on silicon grown by atomic layer deposition from cyclopentadienyl precursor, 84, © 2004 American Institute of Physics.*

In another study [119] metallocene compounds $(MeCp)_2ZrMe_2$ and $(MeCp)_2Zr(OMe)(Me)$ were applied as precursors with water to deposit ZrO_2 at temperatures between 300 and 500°C. Excellent uniformity was gained with both precursors at temperatures between 350 - 400°C. The growth rates were 0.5 Å/cycle. The low concentrations of impurities were found in the films and crystal structure was polycrystalline. The cross-sectional TEM showed that the size of individual crystals was typically between 5 and 15nm, and they often extended through the entire thickness of the film (see Figure 4.17).

As in a study mentioned earlier of TiO_2 [86] , ZrO_2 was also deposited with a heteroleptic precursor and ozone. The novel processes were based on cycloheptatrienyl (CHT, $-C_7H_7$) – cyclopentadienyl (Cp, $-C_5H_5$) compounds of Zr. These precursors provide thermal stability, good reactivity and hence high purity films. The growth rate for CpMeZrCHT and O_3 was 0.7–0.8 Å/cycle for a temperature range of 300–450°C. At 350°C carbon and hydrogen impurities were low; 0.1 and 0.5 at.%, respectively. The growth temperatures over

400°C was reported to cause partial precursor decomposition. The results showed low capacitance equivalent thickness of 0.8.nm and low leakage current density. It seems that heteroleptic compounds offer an interesting group of precursors for further development of group IV metal oxide ALD films.

4.6 Hafnium Dioxide

Hafnium dioxide, HfO_2 has been considered to be one of the most promising applicants to replace SiO_2 gate oxide in complementary metal oxide (CMOS) technology. It has high thermal and chemical stability, a comparatively high permittivity, wide bandgap, and large offsets to silicon.

4.6.1 Processes and Properties of HfO_2

4.6.1.1 The Halide Precursor and Processes

As for many other ALD grown oxides HfO_2 has been deposited with a chloride/water process [120]. The $HfCl_4$ precursor is a solid and has to be heated to 130°C in order to obtain a sufficient vapor pressure. Depositions were done onto soda lime glass at 500°C, and the deposition rate was reported to be 0.5 Å/cycle. Films grown were between 150 and 4000 Å thickness. The films were found to be stochiometric HfO_2 with less than 1% chlorine contamination by Rutherford Back Scattering (RBS). X-ray Diffraction (XRD) studies indicated that growth was initiated with a thin amorphous layer, less than 400Å, followed by polycrystalline growth. In fact, the thinnest film, 150Å, appeared to be essentially amorphous. Atomic force Microscopy (AFM) was used to examine films of different thickness to see how their morphology developed. It was noticed that surface roughening was caused by agglomeration in the early stages of film growth.

In a later study [121] the same process was used to grow films at temperatures from 180 to 600°C. In particular, these investigators were interested in optical quality films with very smooth surfaces. The investigators reported the chlorine contamination to be quite high (~ 5%) at a deposition temperature of 225°C. At 300°C it was about 1%, and at 400°C and above there was none. They further reported that the films less than 250–300 Å thick and grown at temperatures less than 300°C were amorphous. Thicker films were polycrystalline. Consistent with this fact, they confirmed, by Scanning Electron Microscope (SEM), that the surface of a 300Å thick film was very smooth.

As in many ALD processes the nucleation does not occur uniformly and material grows first as islands. This type of initiation causes rough thin films which have poor gate leakage characteristics. In addition for depositions onto bare silicon at high temperatures, the underlying silicon can get rapidly oxidized. To avoid these problems a thin layer of SiO_2 has been grown before depositing a thin film of HfO_2 [122, 123]. In some cases, other surface treatments, such as nitridation [124] of the bare silicon surface, or deposition of an initial very thin layer of Al_2O_3 have been tried [125] to improve results.

The $HfCl_4$ and water process has been a commercial process for many years. Since then many new processes have been scrutinized in order to produce ultra thin and high quality HfO_2 films. Next we discuss the development of a hafnium iodide, HfI_4 process.

Like hafnium chloride hafnium iodide HfI_4 is also a solid that has to be vaporized at temperatures from 130 to 210°C. The high vaporization temperatures make the process very challenging. However, concerning the halide residues in the HfO_2 films, the iodide process might be an attractive choice.

In one study HfO_2 films were deposited from HfI_4 and water at 300°C [126]. Si(100) without the native oxide was used a as substrate material. The growth rate was found to be the same as with the chloride process, 0.5 Å/cycle. Higher growth rate of 0.75 Å/cycle was achieved in another study where H_2O_2 was used as an oxidizer [127]. The growth rate did not vary significantly over a temperature range of 225–500°C. The contamination due to incorporation of iodine into the film was only 0.85% at a deposition temperature of 225°C. This compares to a value of 5% chlorine in the process run with $HfCl_4$ and H_2O [128]. Whether the lower contamination percentage can be attributed to the use of the iodide or to the use of a more effective oxidizer (hydrogen peroxide rather than water) is not clear.

Other experiments were carried out with HfI_4 and O_2 onto Si(100) without native oxide [129, 130]. The pulse and purge times were all 2 s. The growth was stochiometric HfO_2 and the growth rate was found to be 1.1–1.2 Å/cycle between 500 and 750°C. In these experiments the solid HfI_4 was heated to 205°C, a temperature where the growth rates reached their maximum values. All of the films were polycrystalline, and grew a 20Å interfacial layer of SiO_2. The dielectric constant of the as deposited films varied between 10 and 16. Also, an increase in leakage current occurred for films grown at the higher temperatures, and the breakdown strength was reduced. It appears that the enhanced crystallinity occurring at the higher deposition temperatures creates numerous grain boundaries that allow electrical currents to flow more easily.

4.6.1.2 Nitrate Processes

The hafnium nitrate $Hf(NO_3)_4$ precursor is a solid which decomposes readily at moderate temperatures. The first experiments heated the nitrate to 80–85°C, and limited reaction temperatures to 180°C [131]. A typical cycle involved nitrate vapor - nitrogen purge - water vapor pulse - nitrogen purge. All pulses were of 0.6 s length. Deposition was performed onto Si(100) wafers that had been dipped in HF to remove any native oxide. Since all depositions reported were done at only one temperature, 180°C, it was not clear whether there is a usable temperature window for this process. The deposition rate ranged from 3.6 Å/cycle for thin films up to 70Å, to 2.9 Å/cycle for thicker films. These values were much higher than those measured for the processes described earlier. X-Ray Diffraction (XRD) analysis showed that the as deposited films were amorphous, and after 5 min. anneal in nitrogen at 850°C became fully crystalline. Measurements by an atomic force microscope (AFM) showed that the initial films were very smooth. A roughness of 3Å was measured for the initial cycle of the deposited film compared to a value of 2Å for the bare silicon. Other measurements indicated that a thin film covered the entire surface. Therefore, the use of this precursor should enable a smooth and uniform initiation of an ALD film of HfO_2.

In the following study [132] the investigators measured the composition with XPS. The film was found to be substantially oxygen rich even after annealing with O/Hf ratio of 2.76. Electrical measurements were conducted in the next study [133] and they indicated that the dielectric constant for films thinner than 100Å was between 10–12. The breakdown voltage was 6–9 MV/cm, and the leakage current was orders of magnitude less than for SiO_2. The excess oxygen may be the reason for the reasonably low dielectric constant.

4.6.1.3 Organometallic Precursors and Processes

Many organometallic precursors have been tested in HfO_2 process. Similarly as with ZrO_2 precurosrs also three alkylamides exist for HfO_2. Tetrakis(dimethylamino)hafnium, $Hf(NMe_2)_4$;

tetrakis(ethylmethylamino)hafnium, $Hf(NMeEt)_4$; and tetrakis (diethylamino)hafnium, $Hf(NEt_2)_4$. All these precursors are liquids.

Diethyl precursor compound and water was tested in HfO_2 deposition over a temperature range of 250 to 350°C onto Si(100) with native oxide removed [134]. The deposition rate appeared to be 1.2–1.3 Å/cycle when a pulse length of 4 seconds was used at temperatures between 300 and 350°C, and is independent of temperature. This confirmed a usable temperature window for this process. The oxygen to hafnium ratio was reported to be 2, and there appeared to be a small amount of carbon (0.6%) in the film. The films were grown as amorphous and no interfacial layer was found between the silicon and hafnium oxide.

Diethyl precursor and ozone was used in another study [135]. The deposition was carried out onto a silicon wafer with an 8Å SiO_2 layer at termperatures between 150–400°C. The growth rate appeared to be constant within the range of 300–500°C. Small amount of carbon and nitrogen residues were discovered in the films. The roughness was measured to be 2 Å for a film of 25Å thickness. Compared to water process the contamination and the roughness decreased slightly.

Similar films were grown when methyl ethyl precursor ($Hf(NMeEt)_4$) was used with water [136]. The ALD window was found between 200 and 250°C and the films were grown on bare silicon and glass. The films were stoichiometric with with carbon (0.3–0.6%), nitrogen (0.1–0.2%), and hydrogen (2–3%) impurities. The films deposited above 150–175°C were found crystallized.

Later, a Hafniumtetrakis(1-methoxy-2methyl-2propolanate), $Hf(OCMe_2CH_2OMe)_4$ and water -process was studied at the first time [137]. Borosilicate glass and Si(100) were used as substrates and a temperature range of 275–425°C was used. It was found that the precursor was partially decomposing and therefore a self-limiting growth was not reached. The films were crystalline containg monoclinic HfO_2 phase at temperatures above 300–325°C. The refractive index was measured to vary between 1.8 and 2.0. The dielectric constants in $Al/HfO_2/Si$ structures varied between 12 and 17.

Another precursor hafnium alkoxide, Hf(mp)4 (mp = 3-methyl-3-pentoxide, $OC(CH_3)(C_2H_5)_2$) was tried for the first time in ALD HfO_2 process [138]. Water was used as an oxidizer and the growth was found to be self-limiting. The ALD window was discovered between 250 and 350°C where the growth rate was 0.9 Å/cycle. The films were grown as amorphous. The films were found to be stoichiometric with less tha 2 at-% of carbon residues. The *C-V* curve was measured for an Al/HfO_2/p-type Si(001) MIS structure with

10.5 nm thick HfO_2 deposited at 350°C.No hysteresis was observed, and the flat-band voltage was small (–1.2V) indicating negligible amounts of mobile ions and oxide-trapped charges. The relative permittivity was calculated to be 8.6.

In the earlier study for ZrO_2 [118] ,HfO_2 was deposited with metallocenes of $(MeCp)_2HfMe_2$ and $(MeCp)_2Hf(OMe)(Me)$. The growth rate for $(MeCp)_2HfMe_2$ process was between 0.4 and 0.48 Å/cycle for 350 and 400°C, respectively. For $(MeCp)_2Hf(OMe)(Me)$ process the growth rate was 0.5 Å/cycle. The process produced nearly stoichiometric HfO_2 with carbon and hydrogen residues being 0.5 and 1.0 at-%, respectively. All films were polycrystalline as deposited, and possessed a thin interfacial SiO_2 layer. As was mentioned in Section 4.5.1.3 [118] TEM showed that the size of individual crystals was typically between 5 and 15nm, and they often extended through the entire thickness of the film which can be seen in Figure 4.17.

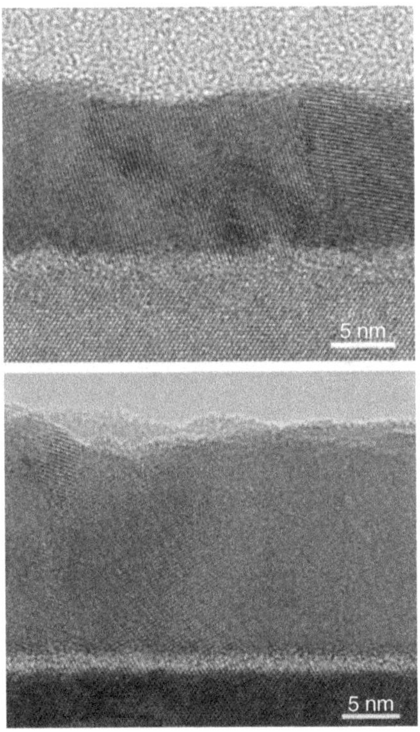

Figure 4.17 Cross sectional HRTEM of a representative HfO_2 film deposited from $(MeCp)_2Hf(OMe)(Me)$ (top) and a ZrO2 film deposited from $(MeCp)_2Zr(OMe)$ (Me) (bottom). Both films were deposited at 350 ° C using 400 deposition cycles [119]. *Reprinted with permission from WILEY-VCH Verlag GmbH & Co.*© 2008.

Cyclopendienyl precursors, [$(MeCp)_2$HfMe(O*i*Pr)] and [$(MeCp)_2$HfMe(mmp)] (mmp= $OCMe_2CH_2OMe$) were synthesized and tested for the first time for HfO_2 ALD deposition [139]. Both precursors evaporate at moderate temperatures which makes them suitable for ALD process. Water was used as an oxidant and a temperature range of 300–450°C was applied. HfO_2 was deposited onto bare Si(100) substrates with a native silicon oxide thickness of 1.5 nm. The films were found to be amorphous. The films contained 0–5 at-% of residual carbon. The dielectric properties were evaluated on the MOS capacitor structure of Al/HfO_2/SiO_2/n-Si. The HfO_2 thicknesses varied between 1.7 - 102 nm. The calculated permittivity was found to be about 14 for [$(MeCp)_2$HfMe(O*i*Pr)] –process and for [$(MeCp)_2$HfMe(mmp)] –process it was about 10. The leakage current density for [$(MeCp)_2$HfMe(O*i*Pr)] –process films was below 10^{-6} A cm^{-2} up to an electric field of 2.5 MV cm^{-1}.

Heteroleptic precursors have already introduced in Section 4.3.1.4 for TiO_2 ,and Section 4.5.1.3. for ZrO_2 . Heteroleptic precursors might be potential candidates for producing high quality ultra thin metal oxide films for CMOS structures. A new heteroleptic hafnium precursor *tert*-butoxytris(ethylmethylamido)hafnium (BTEMAH) and ozone –process was tested to create HfO_2 thin films [140]. The growth rate was found high and the high density HfO_2 films possessed an outstanding thermal stability in electrical testing. The HfO_2 films grown on rutile TiO_2 films achieved a high dielectric constant of 29.

4.7 Other Oxides

Commercial applications of these films have been explored for many years, and include use as transparent heat reflecting films, transparent conducting layers in low cost heterojunction solar cells, protective coatings, gas sensors, and catalysts. During the last decade the number ALD studies have increased in order to explore the growth of these oxides.

4.7.1 Tin Oxide

Another important material is tin dioxide SnO_2 which is used to detect combustible and toxic gases such as CO, NH_3, NO_2, H_2S and CH_4. Initial experiments to grow SnO_2 with ALD [141] were carried out at temperatures from 300–600°C using $SnCl_4$ and H_2O vapor pulses, and nitrogen purges between them. All films were grown on glass. The tin precursor was a liquid, and was maintained at 8°C. That the deposition process was self-limiting was confirmed by demonstrating that the growth rate per cycle was unchanged as the pulse duration for each reactant was increased. Also, film growth as a function of deposition temperature was measured. The ALD cycle timing consisted of $SnCl_4$, H_2O and N_2 purge pulses of 0.2, 0.6 and 3 seconds duration, and it was found that the deposition rate increased from 0.1 Å/sec at 300°C to 0.3 Å/sec at 500°C. Evidently, the reaction rate between $SnCl_4$ and H_2O was too slow at the lower temperatures, contributing to the lower deposition rate. It was also observed that the deposition rate held steady between 500°C and 550°C, so that a useful temperature window was confirmed.

All films grown were polycrystalline, and were close to stochiometric, to within ± 2–3%, and there was no detectable chlorine (<0.2%) as measured by RBS. A depth profile of a 2050 Å thick film by SIMS clarified that there is chlorine content in this film although it could not be detected by RBS. This profile also demonstrates that the Sn, O, and Cl atoms appear to be uniformly distributed with depth for this film.

Experiments were carried out at temperatures from 400–750°C. Deposition rates varied from 0.4 to 2.4 Å/cycle, depending both on the temperature as well as the substrate used (either sapphire or thermally oxidized single crystal silicon). Although there may, indeed, have been self-limited reactions at each temperature, the composition of the chemisorbed iodide precursor must have varied as the temperature was changed. However, it is not clear why the substrate choice should lead to large differences in deposition rates (i.e. 2.4 Å/cycle at 750°C on SiO_2 compared to 1.2 Å/cycle on sapphire), unless it is related to film morphology (roughness).

In another study, film microstructure was studied for films grown from $SnCl_4$ and either H_2O or H_2O_2, or from tin iodide SnI_4 and H_2O_2 or O_2, at the relatively high temperature of 600°C [142]. In general, the films grown from the chloride were rough, and those from the iodide were smooth. Also, the iodide based films were epitaxial when grown on sapphire. When oxygen was used to grow this latter film, it was observed that a 100 Å thick film was sensitive to CO, so that this film could be useful as a sensor.

As noted earlier, the electrical conductivity of SnO_2 films can be increased by adding various dopants. One of the more frequently studied additives is antimony. In a follow up to their earlier paper [143] these authors grew films using a $SbCl_5$ and water cycle interspersed between a large number of $SnCl_4$ and water cycles to grow such doped films. As expected, the resistivity of these films was found to be 0.001 Ω cm.

A water and chloride process has also been used to deposit undoped SnO_2 onto porous silicon [143–145]. An AFM-study was conducted with Sb-doped SnO_2 which was deposited onto soda lime glass, Corning 7059 glass, thermally grown Si (100), and onto sapphire ($1\bar{1}02$). The process was described in earlier studies [143–145].

The effect of film thickness on the electrical and gas-sensing properties of ALD grown SnO_2 was investigated by Natarajan and Cameron [146]. SnO_2 thin films were deposited using $SnCl_4$ and water as precursors and nitrogen as a carrier gas. The films were grown onto Si substrates containing 1 μm thick SiO_2 surface layers at 500°C substrate temperature. O_2 or N_2 were used as carrier and as post-flow gases. The term "post-flow" referred to the gas flow after deposition while the reactor was cooling from growth temperature to room temperature. The chlorine impurities were studied with Rutherford back-scattering spectroscopy (RBS) and the chlorine levels were found to be under the detection limit. The fact that O_2 increases the resistance especially in the oxide films has been a known fact. In this study the resistance increased by 60 times when O_2 was used as a carrier gas and as post-flow (cooling) gas. Compared to regular deposition where N_2 is used as a carrier gas and the films are cooled in a vacuum. The sheet resistance as a function of SnO_2 film thickness was also studied. The results are presented in Figure 4.18.

From Figure 4.18 it can be seen that sheet resistance increases with increasing film thickness up to 50 nm and after that it reaches a plateau. Since oxygen commonly increases the resistance the authors concluded that the occurrence of lower resistance at lower thicknesses is due to the oxygen depletion layer which reaches to a certain critical thickness. Above that thickness there are no changes in the oxygen content and also the resistance stabilizes. In addition gas-sensing functionality of the films was tested with ethanol vapor. As expected it was discovered than the sensitivity depended on the film thickness so that the sensitivity decreased when the film thickness increased. In a subsequent study [147] ALD grown SnO_2 was applied in a microhotplate gas sensor for the first time.

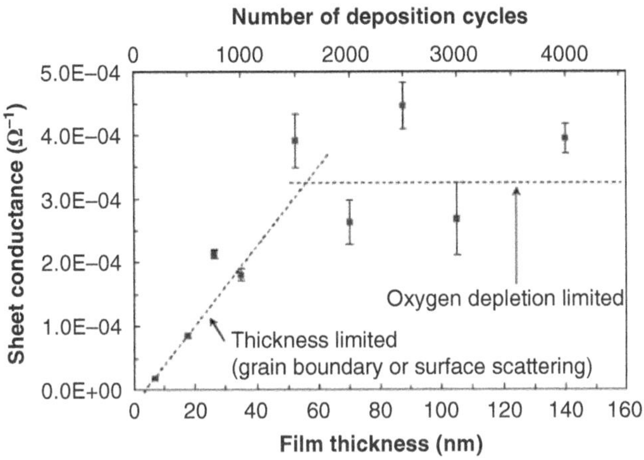

Figure 4.18 Variation in the sheet conductance of SnO_2 films with film thickness. Number of deposition cycles corresponding to the film thickness is given in top *x* axis. The *dotted lines* are given to guide the eyes [146]. *Reprinted with permission from Springer Science: Appl. Phys. Influence of oxygen depletion layer on the properties of tin oxide gas- sensing films fabricated by atomic layer deposition, A 95, 2009, 631, G. Natarajan, D. C. Cameron, © 2009.*

4.7.2 Indium Oxide

Another widely used transparent electrically conducting thin film is indium oxide. Initial studies were done by using $InCl_3$ and water as precursors [148]. The solid indium precursor was heated to 285°C in order to enable sufficient dosing of the substrate surface. Depositions were carried out on soda lime glass at 400–500°C. The maximum growth rates achieved were 0.27 Å/cycle at 500°C. From RBS measurements, the films appeared to be stochiometric with chlorine contamination less than 0.5% and hydrogen content less than 0.2%. They were shown by XRD analysis to be cubic polycrystalline In_2O_3, and exhibited surface roughening for films on the order of 2000 Å thick. Their transparency was over 90%, and the refractive index was 2.05–2.10. Electrically, the films were good conductors (3–6 × 10^{-3} Ω cm).

In order to improve the film conductivity, additional experiments were carried out using tin as the dopant [149] These films are generally referred to as ITO (indium tin oxide). For this process the tin was supplied by $SnCl_3$ vapor. By varying the number of tin cycles versus the number of indium cycles, the tin concentration could be varied from 3–30%. Deposition rates were somewhat lower than for the undoped films (0.20 Å/cycle). For tin content less than 6%, resistivities of 2.4–2.7 × 10^{-4} Ω cm were measured. In the same way as for the undoped films, they exhibited transparency values over 90%.

4.7.3 Tantalum Oxide

Tantalum oxide is another pure metal oxide that can be deposited by thermal ALD at moderate temperatures from a number of precursors. This particular film has been used in the fabrication of capacitors for microprocessor dynamic random access memories (DRAM), because of its high dielectric constant. Initial work was done with tantalum pentachloride, and later studies explored the alkoxide as well as the pentiodide. As before water, hydrogen peroxide or oxygen gas are used as oxidizing agents. Both amorphous as well as polycrystalline layers have been grown.

Deposition of tantalum oxide by thermal ALD can be successfully done using a number of different precursors as shown in the following section. However, there was a tendency for the films to crystallize so that they tended to be electrically leaky. They also contained either halide impurities if deposited from a halide, or carbon and hydrogen impurities if deposited from an organometallic. Although tantalum oxide is a high dielectric constant material (~ 35), films grown on silicon were observed to exhibit much lower values (~ 15). Here as well the radical enhanced ALD processes may bring advantages to film growth.

4.7.3.1 *Films from Tantalum Pentachloride*

Initially, the feasibility of ALD of tantalum oxide from tantalum pentachloride and water was described in a short note almost 25 years ago [150]. In recent years, however, this process has been examined more completely [151], making use of newer analytical tools only recently available. Depositions were done on glass substrates over a temperature range of 300–400°C.The pentachloride is a solid that has an acceptable vapor pressure at 75–90°C, nitrogen was used as a carrier as well as a purge gas.

Experiments were carried out with fixed purge times (0.5,1.7 sec), a fixed water pulse length (0.2 sec), and two different $TaCl_5$ pulse times at different temperatures. At a substrate temperature of 300°C, it was found that the growth rate was 0.39 Å/cycle for a 0.2 sec $TaCl_5$ pulse time, and 0.26 Å for a 1.0 sec $TaCl_5$ pulse time. In other words, the growth rate was reduced as the length of the $TaCl_5$ pulse time was increased. For a fixed $TaCl_5$ pulse time of 0.2 sec, experiments at 300, 325, and 350°C demonstrated a similar phenomenon. Deposition per cycle was found to fall from 0.39Å/cycle to 0.16Å/cycle.

It was also found that films grown at 300°C, or less, tended to be amorphous, while those grown at higher temperatures were at least partially crystalline. One reason for the lower growth rates at higher deposition temperatures and/or longer $TaCl_5$ pulse times, was speculated to be etching of Ta_2O_5 by $TaCl_5$. In order to confirm that this was the mechanism responsible for this behavior, experiments were done where amorphous films were etched in $TaCl_5$ vapor at temperatures from 300 to 450°C. As shown in Figure 4.19, the etch rate increases strongly with temperature. It was also noted that the etch rate of films that were strongly crystalline was about three times slower.

As far as film morphology was concerned, the amorphous films were smooth and featureless. However, the films deposited at higher temperatures than 300°C (or longer $TaCl_5$ pulse times at 300°C), and described as crystalline, had an unusual structure. As can be seen in Figure 4.20, the crystalline films consisted of isolated crystallites embedded in an amorphous matrix. At higher temperatures they are more numerous, and tended to coalesce.

Apparently, as higher temperature films are grown, and crystallites are formed, the amorphous matrix is etched away rapidly, enhancing

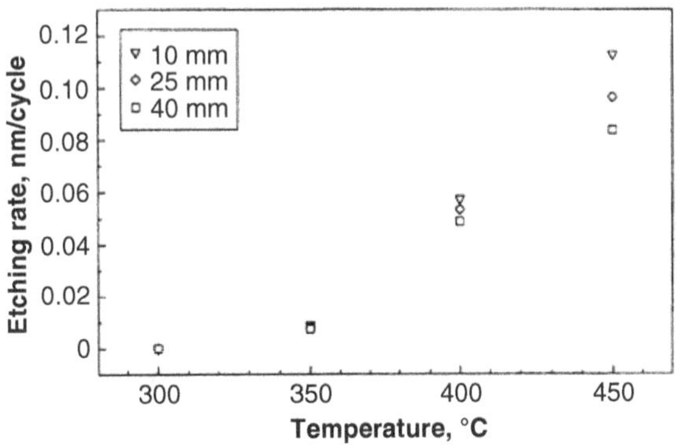

Figure 4.19 Etching rate of amorphous Ta_2O_5 by $TaCl_5$ vapor vs. temperature. Inset identifies the location downstream of the substrate leading edge where measurements were made [151]. *Reprinted from J. Crystal Growth, 212, K. Kukli, M. Ritala, R. Matero, M. Leskala, Influence of atomic layer deposition parameters on the phase content of Ta_2O_5 films, 459, © 2000, with permission from Elsevier.*

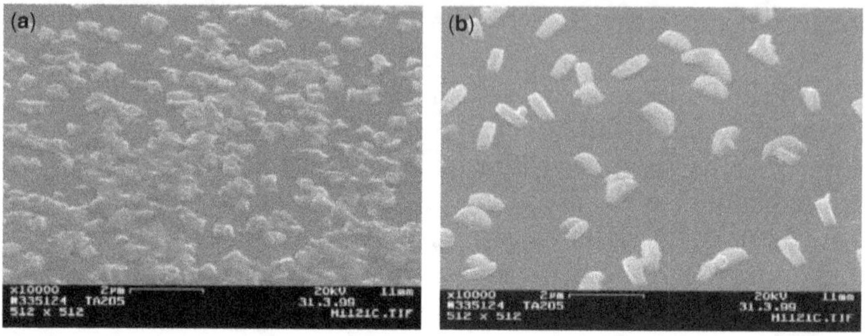

Figure 4.20 Scanning Electron Microscope images of films deposited at (a) 400°C and (b) 350°C on Corning 7059 borosilicate glass [151]. *Reprinted from J. Crystal Growth, 212, K. Kukli, M. Ritala, R. Matero, M. Leskelä, Influence of atomic layer deposition parameters on the phase content of Ta_2O_5 films, 459, © 2000, with permission from Elsevier.*

the polycrystalline nature of the resulting film. This is particularly true because the crystallites do not etch very readily in $TaCl_5$.

The only chemical information reported in [151] was the observation that the film was stoichiometric Ta_2O_5, and that there was chlorine contamination in the film. No measurement of either hydrogen or carbon was reported.

4.7.3.2 *Use of Ethoxide to Grow Tantalum Oxide*

Another well studied ALD deposition process for the growth of Ta_2O_5, makes use of tantalum ethoxide $Ta(OC_2H_5)_5$ and H_2O. Experiments were carried out in the same type of ALD flow reactor used in the previous study [152, 153] at temperatures between 150 and 450°C. Tantalum ethoxide $Ta(OC_2H_5)_5$ is a liquid at room temperature and chemically stable at the temperature necessary (105°C) to heat it to reach the desired vapor pressure. Again, nitrogen was used as a carrier and purge gas. For temperatures between 225 to 325°C a deposition rate of 0.4 Å/cycle independent of temperature was found, as shown in Figure 4.21.

Again, the film composition is stochiometric Ta_2O_5, and all films deposited were amorphous. Hydrogen was detected at a level of 4 at.% at a deposition temperature of 250°C, and at 0.6 at.% at 325°C by nuclear reaction analysis.

From the XPS analysis done it appeared that there was some carbon contamination (~ 2–3 at%), but there was some uncertainty in the measurements.

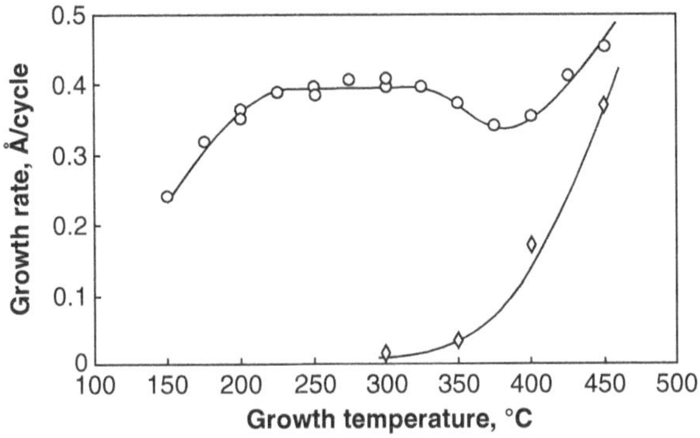

Figure 4.21 Growth rate as a function of substrate temperature. Upper curve ALD, lower curve pyrolysis of $Ta(OC_2H_5)_5$ with no water pulse [152]. *Reprinted from J. Electrochem. Soc. 142, K. Kukli, M. Ritala, M. Leskelä, Atomic Layer Epitaxy Growth of Tantalum Oxide Thin Films from* $Ta(OC_2H_5)_5$ *and* H_2O*, 1670, © 1995, Reproduced by permission of ECS – The Electrochemical Society.*

The refractive index of these films was found to be in the range of 2.14 to 2.23, with the highest value seen for temperatures above 300°C. Also, the dielectric constant was measured to be 21 at 250°C and 25 at 325°C. Both the refractive index and the dielectric constant values were in the same range found for the previous process when $TaCl_5$ was used.

Since one major application of such films is for use as the dielectric in SRAM capacitors, the question of leakage current density has also been studied. For 1500 Å thick films grown on ITO, current voltage curves were generated for films grown at two temperatures. As shown in Figure 4.22, the 325°C film showed the lowest leakage current. However, the leakage current for this film was 2.3 mA at 1MV/cm, and for very thin films this is a higher value than would be desired. The issue of leakage current with high dielectric constant oxide films will be discussed again when we examine the behavior of laminated films later in this Chapter.

Although it is clear that there is a self-limiting mechanism occurring when a substrate is exposed alternately to the ethoxide and water vapor, at appropriate temperatures, so that a valid ALD process occurs, it is interesting to examine the surface chemistry that may be governing [153]. The simplest assumption would be that the alkoxide molecule chemisorbs intact onto a tantalum oxide surface and then reacts with a water molecule in the second step. However, another scenario is more likely.

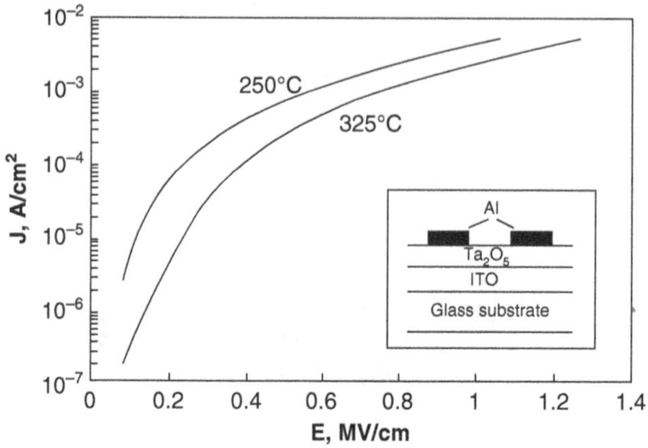

Figure 4.22 Leakage current density vs. electric field for 1500 Å thick Ta_2O_5 films [152]. *Reprinted from J. Electrochem. Soc. 142, K. Kukli, M. Ritala, M. Leskelä, Atomic Layer Epitaxy Growth of Tantalum Oxide Thin Films from $Ta(OC_2H_5)_5$ and H_2O, 1670, © 1995, Reproduced by permission of ECS – The Electrochemical Society.*

$$\text{n(-OH) (ads)} + \text{Ta(OC}_2\text{H}_5)_5 \text{ (gas)} \rightarrow \text{(-O-)}_n\text{Ta(OC}_2\text{H}_5)_{5-n} \text{ (ads)} + \text{nCH}_3\text{CH}_2\text{OH (gas)} \quad (4.10)$$

$$\text{(-O-)}_n\text{Ta(OC}_2\text{H}_5)_{5-n} \text{ (ads)} + \text{(5-n)H}_2\text{O (gas)} \rightarrow \text{(-O-)}_n\text{Ta(OH)}_{5-n} \text{ (ads)} + \text{(5-n)CH}_3\text{CH}_2\text{OH (gas)} \quad (4.11)$$

In other words, the alkoxide reacts with OH radicals on the surface first where it partially decomposes, and then reacts with the water molecule to form Ta_2O_5. If this mechanism is indeed what occurs then this would explain why the use of either water or hydrogen peroxide are such effective oxidizing agents. However, it must be kept in mind that ALD processes for the growth of metal oxides using pure gaseous oxygen have also been demonstrated.

One other process has been reported for the deposition of Ta_2O_5 from the alkoxide [154]. Here ammonia is used instead of water as the second reactant along with argon as a purge gas for deposition on Si(100) wafers. A self limiting growth was demonstrated for temperatures between 250 and 275°C. Since $Ta(OC_2H_5)_5$ contains the necessary oxygen for the growth of the TaO_2 film, the NH_3 apparently served as a catalyst to facilitate decomposition of the alkoxide. The as deposited films were amorphous, and contained nitrogen and carbon impurities

(33% Ta, 51% O, 5% C, and 12% N) as determined by Auger electron spectroscopy. There was no report of any hydrogen impurity.

The film dielectric constant was 13.3, after annealing in oxygen at 800°C. This low a value could have been due to oxidation of the underlying silicon during annealing.

When REALD is used to deposit a tantalum oxide thin film from tantalum ethoxide [$Ta(OC_2H_5)_5$] and oxygen atoms [155] on silicon, in contrast, it is found that stoichiometric Ta_2O_5 is grown with carbon and hydrogen impurities below 0.1%. 150 Å thick films were grown in a capacitively coupled reactor, operated at 13.56 Mhz, and at a deposition temperature of 260 °C. As deposited they were amorphous, and remained amorphous after annealing in oxygen at 700 °C for 2 minutes.

Although an interface layer 17 Å thick on the silicon substrate was formed, the resulting film displayed a dielectric constant of 38, and a leakage current density of 10^{-07} A/cm^2 at 1 MV/cm. Also, these results were largely independent of whether the film was as-deposited or annealed. These measurements have to be compared to the thermal ALD film. From Figure 4.24 we see a reported leakage current of 10^{-03} A/cm^2 at 1 MV/cm, and a dielectric constant of 21–25.

One explanation of the improved characteristics of the REALD films, may be that the interface layer between the silicon and the deposited Ta_2O_5 film may be a tantalum silicate, which would tend to be a higher dielectric constant material than silicon oxide as well as a better quality insulator. Based on these results, additional studies would be justified to clarify the mechanisms involved in the deposition of such REALD based films.

A more recent study on REALD of tantalum oxide film from tantalum ethoxide and oxygen radicals grown at 150 °C and 250 °C without post-annealing reported similar or slightly improved electrical properties. Leakage current density at 1 MV/cm electric field for both films was 10^{-8} A/cm^2, and the dielectric constant for 150 °C grown film was 28 and 36 for 250 °C grown film. These values were clearly improved from the thermally grown films at the same temperature using water as a counter reactant. This study also provided insights into the difficulty in obtaining good electrical properties with the thermal ALD. When a water pulse was added to the REALD cycle by sequences of H_2O-Ta-O or Ta-H_2O-O. The leakage current density and the dielectric constant for the film with water-Ta-O sequence was 4.5×10^{-4} A/cm^2 and 31 respectively. Reduced electrical properties were attributed to the ALD half-reaction where tantalum ethoxide adsorbs on the hydroxyl covered surface. This can be avoided by using oxygen radicals instead [156].

Even if this multilayer approach using both water and radicals to form metal oxide did not result in good film performance in this particular case, it may be useful in some applications. One example is deposition on sensitive substrates, such as polymers. For example, exposing a substrate to TMA and water should form a multilayer, and the energy released from the de-excitation of a noble gas metastable atom should produce a layer of aluminum oxide. Such a process could be particularly useful if thin layers of aluminum oxide needed to be deposited onto plastic substrates, since the carbon in the plastic hydrocarbon could otherwise be attacked. The multilayer approach will also be shortly discussed in Section 4.9.

4.7.3.3 *Films Grown from Tantalum Pentaiodide*

As an alternative to the chloride precursor, the use of tantalum pentaiodide TaI_5 has been studied with an aqueous solution of hydrogen peroxide as the oxidizer [157]. Actually, this tantalum precursor is a solid and requires a substantial temperature to sufficiently vaporize it. For these experiments the precursor was placed in a boat in the reactor chamber and heated to a temperature over 240°C, and nitrogen carrier gas was used to introduce it to the reaction zone. Depositions were carried out onto samples attached to a quartz crystal microbalance. A typical QCM readout is shown in Figure 4.23.

ALD films can be deposited over temperatures from 240–400°C as demonstrated in Figure 4.24. The cycle sequence for these depositions consisted of four steps.

1. TaI_5 pulse, t_1
2. Nitrogen purge, t_2
3. H_2O_2, t_3
4. Nitrogen purge, t_4

As can be seen in Figure 4.24, deposition with a short TaI_5 pulse ($t_1 = 1$ sec) gives a film growth rate of about 0.4 Å which is independent of temperature. For a longer TaI_5 pulse, deposition decreases as deposition temperature increases. The explanation of this behavior is that the TaI_5 is capable of etching the growing Ta_2O_5 layer, while it is being grown, and this capability increases with substrate temperature. It was also found that there was little residual iodine in the deposited films.

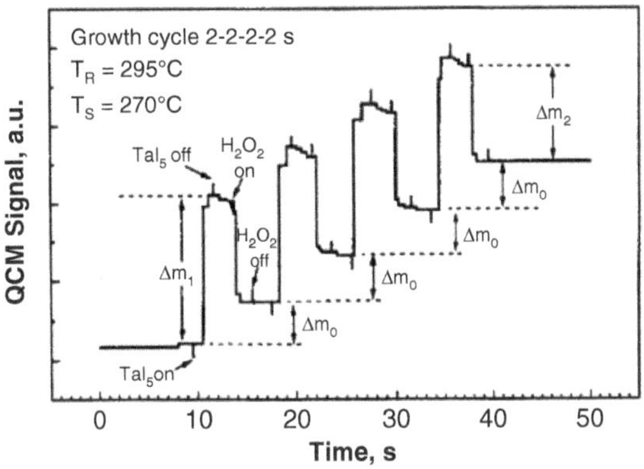

Figure 4.23 Surface mass changes during ALD with TaI_5 and H_2O_2 [157]. *Reprinted with permission from Chem. Mater.,13, K. Kukli, J. Aarik, A. Aidla, K. Forsgren, J. Sundqvist, A. Harsta, T. Uustare, H. Mandar, and A-A. Kiisler, Tantalum Oxide Thin Films from Iodide Precursor, 122, © 2001 American Chemical Society.*

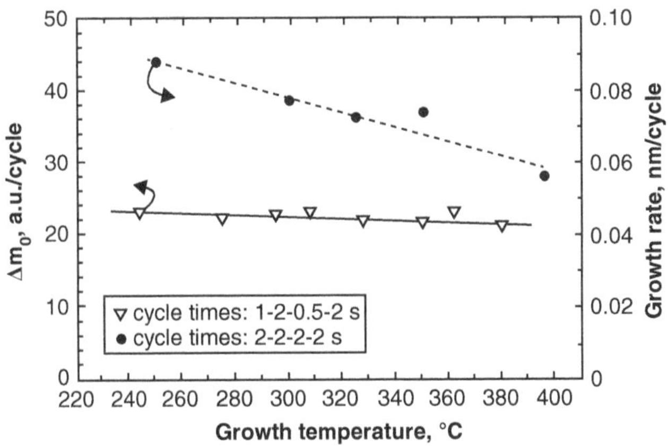

Figure 4.24 Growth rate dependence on growth temperature for 1000 growth cycles [157]. *Reprinted with permission from Chem. Mater.,13, K. Kukli, J. Aarik, A. Aidla, K. Forsgren, J. Sundqvist, A. Harsta, T. Uustare, H. Mandar, and A-A. Kiisler, Tantalum Oxide Thin Films from Iodide Precursor, 122 © 2001 American Chemical Society.*

In fact, they were stochiometric in composition. Films grown onto fused silica were all amorphous, while those grown on Si (100) were amorphous at deposition temperatures below 350°C, but partially crystallized at higher temperatures.

In general, these films were similar to films grown from $TaCl_5$, except that they tended to have a lower level of residual halide contamination. They also tended to be more amorphous. Of course, the practical difficulties in building an ALD reactor that uses a solid precursor that has to vaporized at such high temperatures should not be underestimated.

Finally, as before, one can use oxygen as the oxidizing agent with TaI_5 [158]. In this case, depositions were carried out at higher temperatures, between 400 and 700°C on Si (100) wafers. The films were iodine free above 450°C, and were polycrystalline Ta_2O_5. Apparently, the growth rate varied strongly with temperature, thereby not offering the typical ALD temperature window.

4.7.3.4 Other Precursors to Grow Tantalum Oxide

A thermally stable pyrazolate-based precursor, Ta(NtBu)-(tBu2pz)3 (tBu2pz) 3,5-di-*tert*-butylpyrazolate) and ozone were used to deposit conformal Ta_2O_5 films [159]. An ALD window was found in the range of 300–450°C. The growth rate was 0.3 Å/cycle with a tantalum-precursor pulse > 1 s. Time-of-flight-elastic recoil detection analysis (TOF-ERDA) was performed to study the film composition. The carbon residue was found to be < 1at-%, hydrogen level was below the detection limit for films deposited ≥375°C but showed values of 13 and 1.9 at-% at 275°C and 325°C, respectively. Nitrogen concentrations were <0.5% at temperatures of 375°C, but increased to 0.9 and 1.8% at 425°C and 475°C.

4.8 Mixed Oxides and Nanolaminates

A large number of simple oxides can be successfully deposited by thermal ALD processes. Using the same concepts, it may also be possible to deposit ternary oxides as either mixed oxides or nanolaminates. By a mixed oxide we are referring to a uniform mixture of two metals and oxygen. A nanolaminate will be a thin film which is composed of alternate layers of two binary oxides. For very thin nanolaminates the alternate layers may be only several angstroms in thickness.

The first issue in forming such films is whether or not there is a thermal ALD process for each metal oxide. Once that has been established, then we need to know if their temperature windows have sufficient overlap to allow us to run both processes at the same temperature. The same question applies to either a mixed oxide or a nanolaminate.

An alternate technique is to find a single precursor that contains both metal atoms. In general, this is quite difficult to do. There are a few examples of the use of this approach, which we will mention later in this chapter. However, as will be seen, the film stochiometry is constrained by the composition of the precursor.

In any case, it is possible to grow some well known, and commercially important ternary oxides, that are difficult to fabricate with perfect conformality and excellent uniformity by other thin film deposition techniques. It also may be possible to fabricate newer materials to achieve unusual characteristics. This chapter will review several processes that will serve to illustrate how mixed oxides and/or nanolaminates may be fabricated.

4.8.1 Mixed Oxide Processes

Several of the ternary mixed oxides that have been explored extensively for commercial applications are high permittivity ferroelectrics such as strontium [160–162] and barium titanate [161], and strontium tantalate [163]. Some success has also been reported in growing lanthanum aluminate [164, 165]. In addition, hafnium aluminate [166–168] has been studied for application as a gate dielectric. Finally, hafnium silicate [169–173], zirconium silicate [174–176], and bismuth silicate [177] have all been deposited by thermal ALD. In this section we will review all of these processes as well as the film quality of the materials.

Development of a thermal ALD process for these ferroelectric films has been difficult because of the lack of suitable Ba and Sr precursors. As has been pointed out earlier in this text many times, in order to create a true ALD process each precursor has to be stable on the surface at the appropriate temperature. In other words, it should not decompose spontaneously at temperatures necessary for the desired surface reaction to proceed rapidly. The obvious precursors would be the β-diketonate compounds of Sr and Ba. However, these compounds will only react with water vapor or oxygen at temperatures so high that they would decompose when striking the surface.

However, thermal ALD of these titanates has been demonstrated [160, 161] using two new compounds: strontium bis(triisopropyl cyclopentadienyl) and barium bis(pentamethylcyclopentadienyl). The titanium precursor was titanium tetraisopropoxide, and water was used as the oxidizer. Using these materials, thermal ALD of the titanates has been shown to be feasible at temperatures in the range of 250–325°C.

The film composition was controlled by alternating Sr-O and Ti-O depositions. In fact, it was found possible to vary the film stochiometry by changing the number of Si-O vs. Ti-O cycles. This is seen in Figure 4.25, where it is interesting to note that when the Sr/Ti-cycle ratio was 1 or above, there was always an excess of Sr in the film. To achieve a 1:1 ratio of Sr to Ti (as appropriate for stochiometric $SrTiO_3$) it was necessary to keep the Sr/Ti cycle ratio at about 0.83. This would correspond to 5 cycles of Sr-O for every 6 cycles of Ti-O.

The data in Figure 4.25 were generated at 325°C. At higher temperatures it was found that the titanium tetraisopropoxide decomposed on the surface before the Sr precursor arrived. At lower temperatures the deposition rate was lower. For example, at 250°C the deposition rate was approximately 0.5Å/cycle compared to values well in excess of 1.0Å/cycle at 325°C.

The only impurities found in these films were carbon (0.1–0.3 atomic %), and hydrogen (0.4–1.2 atomic %). Also, the crystallinity was dependent on the deposition temperature and the water dose per cycle.

For low water dosage per cycle (18 μg), the lowest temperature (250°C) produced amorphous films. As either the temperature or the water dosage was increased the films became increasingly crystalline. Annealing the amorphous films for an hour in air at 500°C produced crystalline films. The same annealing process applied to those films that were partially crystallized improved their crystal quality. For example, the permittivity of 2800Å thick films formed at 325°Cwas quite high, as is shown in Figure 4.26.

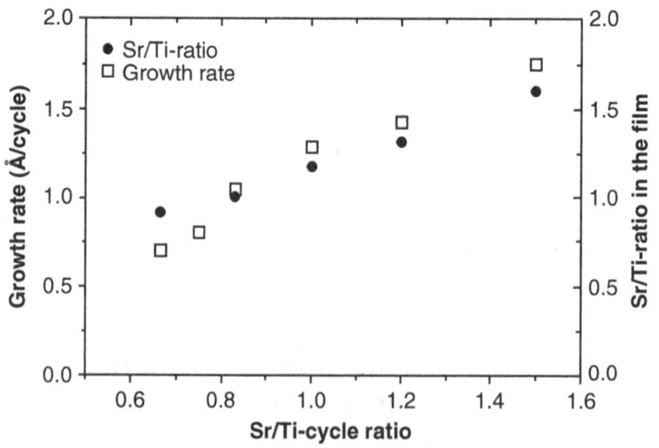

Figure 4.25 Growth rate and composition of $SrTiO_3$ films as a function of the (Sr-O)/(Ti-O) cycle ratio at 325 °C [160]. *Reprinted with permission from Wiley- VCH Verlag GmbH & Co, © 2001.*

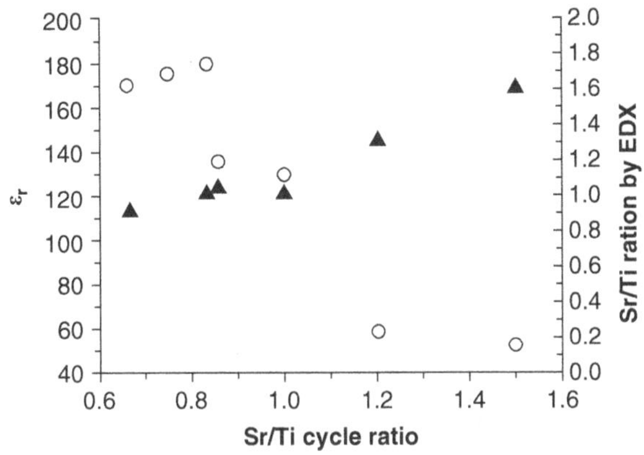

Figure 4.26 Permittivity vs. Sr/Ti ratio of strontium titanate films [161]. Note that ▲ represent the Sr/Ti ratio, and o represent ε_r. *Reprinted from Electrochem and Solid State Ltrs.2, M. Vehkamäki, T. Hatanpää, T.Hänninen, M. Ritala, M. Leskelä, Growth of $SrTiO_3$ and $BaTiO_3$ Thin Films by Atomic Layer Deposition,504, © 1999, Reproduced by permission of ECS – The Electrochemical Society.*

Similar results were found for barium titanate, except that this process was less reproducible in terms of film composition and permittivity.

Another mixed oxide, that is being considered for use in the development of a capacitor dielectric for a nonvolatile random access memory, is the ferroelectric strontium tantalate ($SrTa_2O_6$). A novel "single source" precursor has been used to deposit this material by thermal ALD with water as the oxidizer [163]. This precursor already has the strontium and tantalum in the correct 1:2 ratio, so that growing stochiometric films may be easier to accomplish. The precursor is strontium bis(tantalum pentaethoxide dimethylaminoethoxide) or $SrTa_2(OEt)_{10}(dmae)_2$.

For the ALD process, the precursor was evaporated, and was stable, at 130°C. Growth temperatures from 200 to 350°C were tried, and as well the precursor pulse length was varied from 0.2 to 3 sec. Nitrogen purging was used between the organometallic precursor and water pulses. The growth rate from 200 to 300°C is constant at 0.24 Å/cycle, and only increases a small amount at 350°C to about 0.27Å/cycle. Therefore, this process appears to have a robust temperature window. When the precursor pulse length is increased at any given deposition temperature, there is some increase in the deposition rate, as shown in Figure 4.27.

The increase in growth rate with the increase of precursor pulse length for depositions at 350°C is indicative of some degree of decomposition of this unique precursor. This may make it difficult to achieve very uniform films on large substrates at this temperature. For temperatures of 300°C, or lower, it should be possible to grow good quality films.

As far as film composition and impurity levels are concerned, we can see from Table 4.1, that inspite of the precursor having a Sr/Ta ratio of 0.50 not every film grown had this ratio. Films deposited at 250°C and 350°C showed Sr/Ta very close to 0.50. However, the 250°C film had a high level of impurities, and the 350°C film exhibited a rapid increase in growth rate with increasing precursor pulse rates as noted in Figure 4.27.

The films, as grown, were all amorphous. Annealing in air at 500°C for one hour gave no evidence of crystallization. However, annealing at 800°C produced fully crystallized material. The permittivity was 16 as deposited, 23 after a 500°C anneal, and up to 50 with the 800°C anneal.

When these films, as well as the strontium titanate discussed in the previous section, are used in thin film capacitors after annealing, the leakage currents tend to be unacceptably high. There are probably many possible reasons for this behavior, but the obvious ones are grain boundaries that must exist in the polycrystalline films.

In the development of high temperature semiconductors, it would be highly desirable to be able to deposit high quality epitaxial

Table 4.1 Sr/Ta ratios and impurities in films deposited with $SrTa_2(OEt)_{10}(dmae)_2$ [163]. *Reprinted from J. Electrochem. Soc. 151, M. Vehkamäki, M. Ritala, M. Leskelä, A. C. Jones, H. O. Davies, T. Sajavaara, E. Rauhala, Atomic Layer Deposition of Strontium Tantalate Thin Films from Bimetallic Precursors and Water F69, © 2004, Reproduced by permission of ECS – The Electrochemical Society.*

Sample pulse time /growth temperature	Sr/Ta ratio	H (atom %)	C (atom %)	N (atom %)
0.6 s/200°C	0.63	3 ± 1	0.5 ± 0.1	0.2 ± 0.1
0.6 s/250°C	0.51	20 ± 4	4 ± 0.5	0.9 ± 0.2
0.6 s/300°C	0.58	3 ± 1	0.7 ± 0.1	0.2 ± 0.1
0.6 s/350°C	0.50	1 ± 0.5	<0.1	<0.1
0.6 s/250°C[a]	0.48	9 ± 1	2.7 ± 0.5	0.6 ± 0.1
0.6 s/250°C[b]	0.53	11 ± 2	3.2 ± 0.5	0.7 ± 0.2

[a] Large water dose.

[b] 2.0 s $SrTa_2(OEt)_{10}(dmae)_2$ purging time.

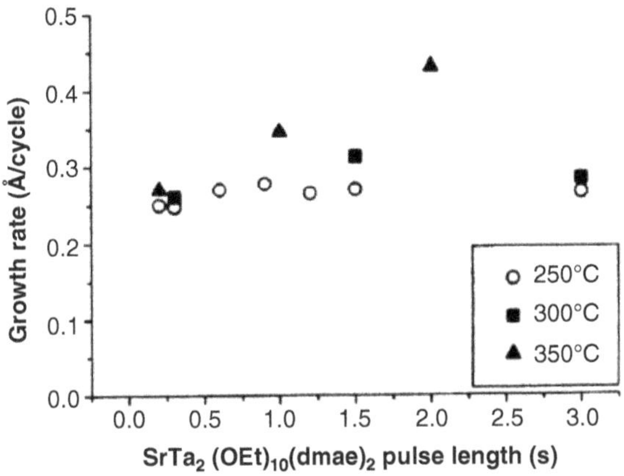

Figure 4.27 Growth rate of Sr-Ta-O films as a function of precursor pulse length [163]. *Reprinted from J. Electrochem. Soc. 151, M. Vehkamäki, M. Ritala, M. Leskelä, A. C. Jones, H. O. Davies, T. Sajavaara, E. Rauhala, Atomic Layer Deposition of Strontium Tantalate Thin Films from Bimetallic Precursors and Water F69, © 2004, Reproduced by permission of ECS – The Electrochemical Society.*

films directly onto single crystal silicon substrates. Unfortunately, problems of interdiffusion, thermal mismatch and differing lattice constants make this very difficult. For this reason, development of a variety of buffer layers has been explored, and lanthanum aluminate ($LaAlO_3$) appears to be one of the better candidates. A process for growth of this material on silicon will be reviewed in this section [164].

A thermal ALD process was used to deposit $LaAlO_3$ from $La(thd)_3$ (thd= 2,2,6,6-tetramethylheptane-3,5-dionate) and $Al(acac)_3$ (acac= pentane-2,4-dionate) precursors with ozone as the oxidizer. The reactor was a flowing system with nitrogen used as a purge gas between precursor exposures. Metal precursor pulse times were 0.8–1.5 sec, the ozone pulse time was 2 sec, and the nitrogen purge times were 1–3 sec. The temperature range for deposition was 325–400°C. In the range of temperatures between 350°C and 375°C, growth rates were between 0.36Å and 0.39Å, which indicates a reasonable temperature window.

The film composition was varied by changing the relative number of La and Al cycles. Results showed that the La to Al ratio could be varied over a wide range. Stoichiometric $LaAlO_3$ films could be grown using optimum pulsing ratios between 1.75 and 2.0. For these films, the impurities consisted of 0.8–1.9 atom% of carbon and 0.3 atom% of hydrogen.

All as deposited films were amorphous. Films deposited onto silicon (100) could be crystallized by annealing in nitrogen at temperatures above 900°C. The annealed films grew columnar grains.

However, epitaxial films could be grown on either MgO-buffered Si(100) or $SrTiO_3$(100). A similar process using La $[N(SiMe_3)_2]_3$ with TMA and water has also been explored [165].

In recent studies [166–168], thermal ALD films of hafnium aluminate with different Hf/Al ratios were deposited from $HfCl_4$ and TMA with H_2O as the oxidant. The films were grown by alternating HfO_2 and Al_2O_3 depositions. In one case [159], the deposition temperature was reported to be 300°C. Although the deposition temperature was not reported in the other study [167, 168], it can be assumed to be similar. The ratio of Al to Hf was varied by adding one or more cycles of either the HfO_2 or Al_2O_3 pulses.

Since one of the motivations for these investigations was to explore this material as an ultra-thin high k gate dielectric, an important feature evaluated was film morphology. As we had seen earlier, Al_2O_3 films were always amorphous at the temperatures used for deposition or annealing. On the other hand, HfO_2 films tended to be amorphous as deposited, but became crystalline at annealing temperatures used in the complete integrated circuit process fabrication. Therefore, one would expect that films with substantial fractions of aluminum would remain amorphous on annealing. The results reported quantified this phenomena. In the first study [166], hafnium aluminate films that had 50% aluminum were polycrystalline after a 1050°C anneal in nitrogen for 5 sec. When the aluminum content exceeded 70%, they remained amorphous for the same annealing condition. In the second study [167], it was reported that the hafnium aluminate remained amorphous with a 74% aluminum content after a 900°C vacuum anneal for 30 min, although there was some evidence of scattered crystal formation. When the vacuum anneal temperature was lowered to 850°C, then films with 34% aluminum also remained largely amorphous.

Then, as demonstrated earlier in this chapter, depositing these 74% aluminum films onto bare silicon wafers caused a fairly thick interlayer (IL) of silicon oxide to grow between the hafnium aluminate and the silicon. One was hafnium rich (Al_2O_3 34%), and the other was aluminum rich (Al_2O_3 74%) [168].

The hafnium rich film experiences a substantial IL growth on annealing. The aluminum rich film also demonstrates IL growth, but to a lesser degree. The IL will degrade the overall dielectric constant of the dielectric stack, since the silicon oxide layer has a fairly small value of ε compared to that of the aluminate.

Again, there was chlorine contamination in these films, but no measurements were reported. However, it was noted that the chlorine content was reduced on annealing in the hafnium rich film. This was probably due to the fact that these films became polycrystalline on annealing, and the grain boundaries provided a preferred diffusion path for out diffusion of chlorine.

In summary, it has been shown that an amorphous layer of aluminum rich hafnium aluminate can be grown by thermal ALD, and will remain amorphous even after high temperature annealing. Much additional work will be necessary to determine if such a film could form the basis of a high k gate oxide.

A number of studies in recent years have explored the feasibility of depositing mixed oxides including silicon, in other words metal silicates. The two mixed oxides of primary interest, because of their possible application as gate oxides for integrated circuits, are hafnium and zirconium silicate. The former will be reviewed in this section [169–173].

In addition to the obvious halide precursors, there are many complex organometallic molecules that have been explored. For example, deposition of hafnium silicate from alternating exposures of tetrakis(dimethylamino)-hafnium, $Hf(N(CH_3)_2)_4$, and tris(tert-butoxy)silanol, $(t\text{-}BuO)_3SiOH$, at substrate temperatures of 250–350°C has been reported [169]. The films were described as having silicon to hafnium ratios between 2–3, so they were very silicon rich. It is not clear if it would be possible to deposit films with higher hafnium content using these precursors. The only impurities reported were carbon and nitrogen at levels less than 2%. However, no data was reported on the temperature dependence of the deposition rate in these experiments, so it is not clear if an acceptable temperature window exists.

Finally, although one rational for the use of these precursors is the avoidance of an interface layer (IL) of silicon dioxide when depositing onto silicon, there is no data reported as to whether or not some IL is formed. Also an important question would be whether such a layer would form, or grow thicker, when these films are annealed at high temperature.

Another novel technique for growing a mixed oxide would be to inject the Si and Hf precursors simultaneously [170], with ozone injected in a second exposure. The two liquid precursors used were tetrakis(ethylmethylamino)hafnium, and tetrakis(ethylmethylamino)silicon.

Since these compounds are chemically compatible, there was no difficulty in injecting their vapor simultaneously. The procedure followed is shown schematically in Figure 4.28 below.

By adjusting the precursor flows these authors demonstrated that film composition may be readily modified, as shown in Figure 4.29. Of course, similar composition control could also be achieved by simply alternating the precursors, since generally much less than a monolayer of film is deposited at each cycle. In fact, this same group also reported on results in a batch reactor with these same reactants injected alternately [171]. However, injecting a mixture may simplify the ALD reactor. Finally, the variation of deposition rate with temperature is shown in Figure 4.30. At the lower temperatures (300–400°C) it appears that there is a 5% variation in deposition rate. Therefore, a true temperature window does not appear to be achieved. Also, it is not clear what film composition this curve applies to, or whether it is the same for all temperatures.

The films appear to be relatively pure with the level of carbon contamination at or less than 1%.

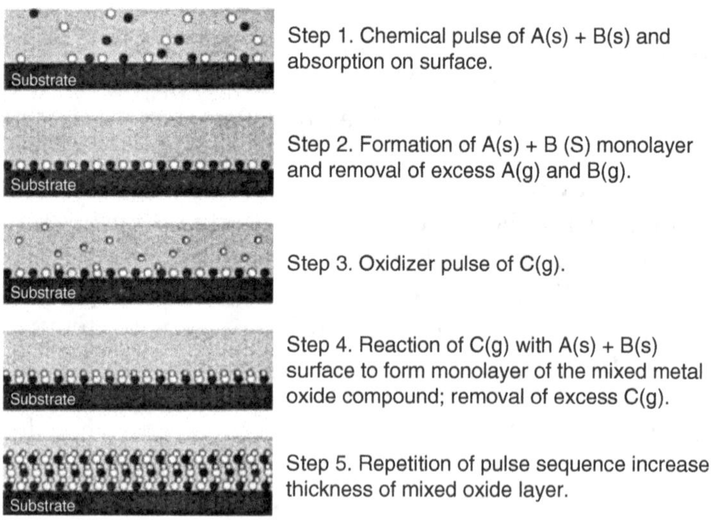

Figure 4.28 Illustration of the co-injection procedure for ALD growth of mixed oxide [170]. *Reprinted with permission from J. Vac. Sci. Technol. A22, Y. Senzaki, S. Park, H. Chatham, and L. Bartholomew, Atomic layer deposition of hafnium oxide and hafnium silicate thin films using liquid precursors and ozone, 1175 © 2004, American Vacuum Society.*

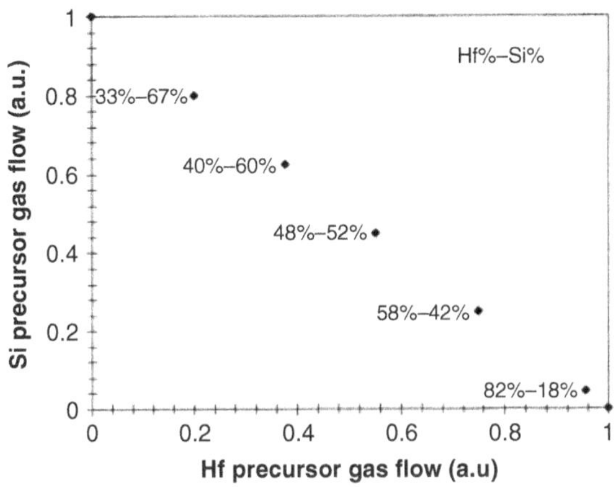

Figure 4.29 Hafnium silicate composition as determined by precursor flows [170]. *Reprinted with permission from J. Vac. Sci. Technol. A22, Y. Senzaki, S. Park, H. Chatham, and L. Bartholomew, Atomic layer deposition of hafnium oxide and hafnium silicate thin films using liquid precursors and ozone, 1175 © 2004, American Vacuum Society.*

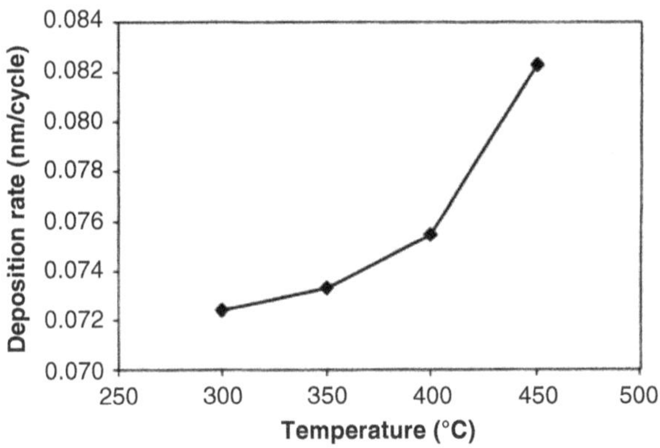

Figure 4.30 Deposition rate versus temperature for $Hf_xSi_{1-x}O_2$ films [170]. *Reprinted with permission from J. Vac. Sci. Technol. A22, Y. Senzaki, S. Park, H. Chatham, and L. Bartholomew, Atomic layer deposition of hafnium oxide and hafnium silicate thin films using liquid precursors and ozone, 1175 © 2004, American Vacuum Society.*

Next, hafnium silicate films can also be deposited from one precursor that contains both Hf and Si, and an oxidizer. In one study [172, 173], dichlorobis [bis(trimethylsilyl)amido]hafnium $HfCl_2[N(SiMe_3)_2]_2$ vapor was used with water as the oxidant. Depositions were carried out from 150–400°C. Although it is claimed that depositions at all temperatures studied were self-limiting, the film deposition rate varied from 1.3 Å/cycle to 0.25 Å/cycle over the temperature range studied. In fact, the only practical temperature window would be between 200–250°C. It was reported, however, that any carbon, nitrogen or chlorine impurities were below 1%. The film composition did vary with deposition temperature, as can be seen in Figure 4.31. However, within the "temperature window" cited above, the film contains mostly hafnium and relatively little silicon.

Finally, another approach to the deposition of hafnium and zirconium silicates is to use one chloride precursor, and a second organometallic precursor. Several approaches to the use of this type of chemistry to deposit zirconium silicate will be discussed in the next section.

Another film of great interest is zirconium silicate, which is similar to the hafnium silicate just discussed, but may be more resistant to diffusion of impurities through it. Of course, the hope is also that such a film could be grown on bare silicon, without creating a layer of low dielectric constant silicon dioxide.

If we work with two separate precursors, and one of them contains oxygen, then it may not be necessary to use a separate oxidant, such as water, to provide the desired oxygen. Two such investigations have been reported. In the first, $ZrCl_4$ is used to contribute the zirconium, and the second precursor is tetra-n butyl orthosilicate (TBOS) $[Si(O^nBu)_4]$ which contributes oxygen and silicon [174].

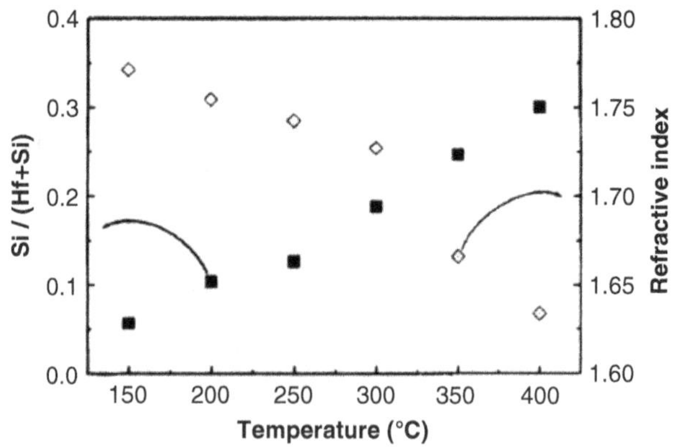

Figure 4.31 Influence of deposition temperature on composition and refractive index for hafnium silicate films [172]. *Reprinted from Electrochem. and Solid State Ltrs. 7, W-H. Nam, S-W. Rhee, Atomic Layer Deposition of Hafnium Silicate Thin Films Using $HfCl_2[N(SiMe_3)_2]_2$ and H_2O, C55, © 2004, Reproduced by permission of ECS – The Electrochemical Society.*

The growth rate of this silicate was stabilized, at a deposition temperature of 400°C, for a $ZrCl_4$ injection time of 5 sec, a 2 sec argon purge, a 2 sec TBOS injection time, and a second 2 sec purge time. Over the the temperature range of 300–500°C, these parameters gave growth rates from 1.5 to 1.1 Å/cycle. Between 350 and 400°C the deposition rate was relatively constant at 1.3 Å/cycle, so a temperature window was found. However, the film composition was largely silicon, and the carbon and chlorine impurities were significant at the deposition temperature of 400°C.

Rather than using the organometallic precursor to contribute silicon, another approach would be to use $SiCl_4$ first and then a zirconium organometallic containing oxygen [175]. For this study, zirconium o-*tert*-butyl [$Zr(O^tC_4H_9)_4$] was chosen. Because these two reactants were substantially more reactive, similar or higher deposition rates were observed at lower temperatures. At temperatures from 175–200°C, the deposition rate was 1.6 to 1.8 Å/cycle, although the variation with temperature was significant. However, with this process, it was possible to deposit films that contained significantly more zirconium than silicon (i.e. Zr/(Zr + Si) = 0.7). It also was reported that the level of chlorine contamination was reduced to a value below 1 %, and that no carbon was observed.

In contrast to the first study cited (ref. 15), another investigation used a different organometallic precursor with the $ZrCl_4$, which was 3-aminopropyltriethoxy silane $NH_2(CH_2)_3Si(OC_2H_5)_3$ (APTES) [176]. In this case, it was possible to deposit films at 300°C with considerably more zirconium (Zr:Si= 2:1), and the chlorine content was below 1 %. Also, it was reported that the films deposited were amorphous as deposited, and did not begin to crystallize until annealed at 1000°C. The film deposition rate was given as 1.2 Å/cycle. Finally, the dielectric constant of this film was somewhere between that of SiO_2 and ZrO_2 (~ 7–14).

As we had seen in earlier section, mixed oxides can sometimes be deposited from precursors that contain both metal (or semiconductor) elements. Another example is found in the ALD deposition of bismuth silicate [177]. In this case, a new precursor tris(trimethylsilylmethyl)bismuth, $Bi(CH_2SiMe_3)_3$ was fabricated. We also recall that the concern when using such precursors arises because of the possible difficulty in controlling the Bi/Si ratio. In this case, we start out with a precursor that has an excess of silicon.

Three bismuth silicate stochiometries are of interest for electronic or optical applications [177]. They are Bi_2SiO_5, $Bi_4Si_3O_{12}$ and $Bi_{12}SiO_{20}$. As can be seen, the control of the Bi/Si ratio of the film being deposited is an important objective of the deposition process.

Films were deposited using ozone as an oxidant over a temperature range from 200°C to 450°C. For these temperatures the films were amorphous, as deposited. A temperature window was found from 250°C to 350°C, where the deposition rate was constant at 0.40 Å/cycle. Over this restricted temperature range the Bi/Si ratio was found to be ~ 2. Although annealing these amorphous films at elevated temperatures gave final films with different Bi/Si ratios, there appear to be many complex issues that need to be studied before this phenomenon is fully understood.

In the same way that we saw that mixed oxides could be deposited by thermal ALD, we can also expect that the same thing can be done with REALD. So far, there are only a few published reports of such film growth, but it is reasonable to anticipate many more in the future.

One film being explored [178, 179] is the high dielectric constant material strontium tantalate or STO ($SrTa_2O_6$) deposited by REALD. The precursor is the bimetallic Sr $[Ta(OEt)_5dmae]_2$ which contains not only the strontium and tantalum in STO but the oxygen as well. The second reactant is atomic oxygen created with an oxygen plasma.

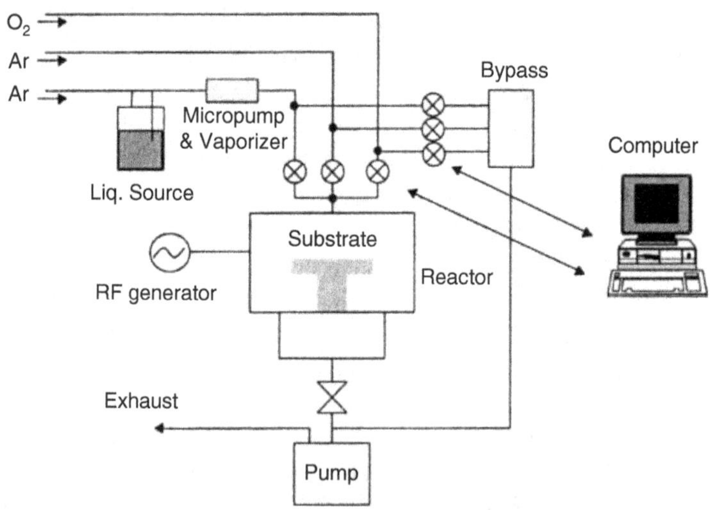

Figure 4.32 Capacitive REALD reactor for deposition of STO [178]. © *2001 The Japan Society of Applied Physics.*

Deposition was carried out in a capacitively coupled RF reactor, with the substrate in contact with the glow discharge, shown schematically in Figure 4.32.

The ALD timing cycle consisted of a pulse of SBT (3 sec), followed by an argon purge pulse (2 sec), followed by an oxygen plasma exposure (2 sec), and the another argon purge pulse (2sec). In order to verify that the process is behaving as an ALD process, The SBT pulse and the deposition temperature were varied, and the thickness per cycle measured. Figure 4.33 shows the results.

As can be seen, for SBT pulses longer than 2 seconds, the deposition rate per cycle is 0.8 Å/cycle, and doesn't vary with temperature in the range of 200–350°C. As expected, the deposited films were conformal.

The as deposited films stochiometry was evaluated by RBS analysis. For ideal STO we would have $SrTa_2O_6$ so that Ta/Sr = 2, and O/Ta would be 3. Measured composition of the deposited film was Ta/Sr = 3.7, and O/Ta was 2.7. Therefore, these films were substantially Sr deficient, as well as slightly oxygen deficient.

The grown films were polycrystalline, both as deposited and when annealed at temperatures from 500 to 800°C. However, the average grain size increased with annealing temperature.

Electrical properties of these films were explored by growing the structure Pt/STO/Pt, and annealing it at the temperatures just

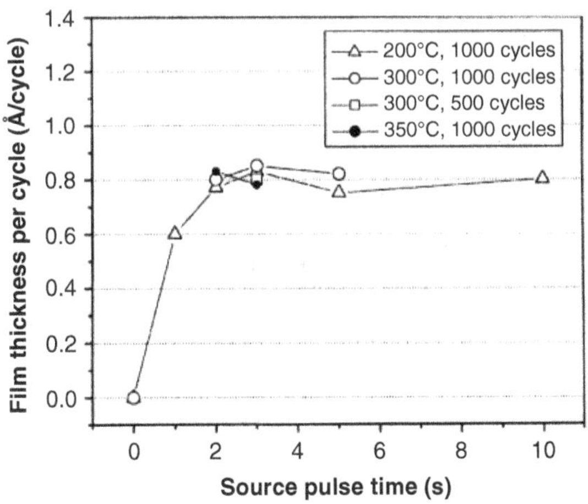

Figure 3.33 Deposition per cycle vs. SBT pulse time and substrate temperature [178]. © *2001 The Japan Society of Applied Physics.*

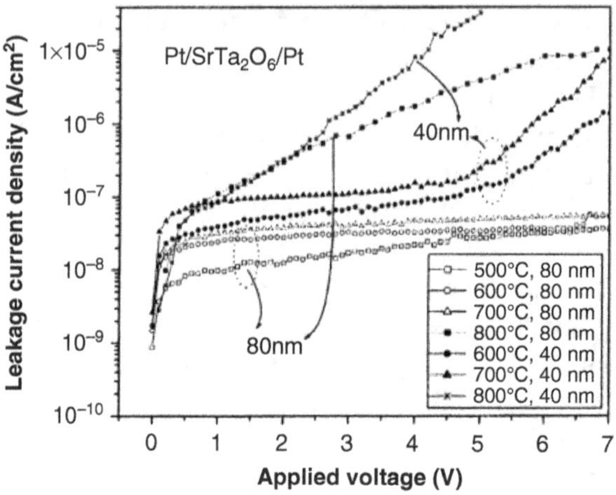

Figure 4.34 IV characteristics of annealed STO films [178]. © *2001 The Japan Society of Applied Physics.*

described. The current-voltage characteristics were measured, and are shown in Figure 4.34.

As long as the annealing temperature is less than 700°C, the leakage current will be under 10^{-07} A/cm^2 for voltages up to 3V. At the same time the dielectric constant of these films is quite high. For an annealing temperature of 600°C it is 40, and at 700°C it is 50.

In all of the above discussion, we were evaluating the dielectric properties of STO deposited onto platinum. However, if such a film were to be considered as a replacement for native SiO_2 in silicon integrated circuits, then we have to evaluate the behavior of a very thin layer deposited onto bare silicon. As we have seen earlier, such deposition always leads to an interfacial oxide layer containing silicon. This layer is typically 20 Å thick, and therefore plays an important role in the application of these films to integrated circuits. A discussion of this issue is given in a subsequent paper for the STO films, and the electrical behavior is encouraging [179].

As we had noted above, the stochiometry of the STO film was dictated by the fact that a single precursor (containing both Sr and Ta) was used. Another, more flexible approach would be to use a separate precursor for each metal. In this way the stochiometry can be controlled by adjusting the number of ALD cycles for each precursor. An example of such an approach [180] is the REALD growth of $SrTiO_3$ from $Sr(METHD)_2$ and $Ti(O^iPr)_4$ with an oxygen plasma. The reactor was a downstream plasma setup, with the plasma generated in a microwave discharge. In order to quantify a temperature window, the deposition of TiO_2 and a SrO depositions were individually carried out. It turns out that between 150 and 275°C, both films deposit at a constant rate of 0.4 Å/cycle. Another interesting feature of this process is the fact that the authors were able to vary the Sr/Ti ratio with deposition temperature, from 2.0 at 150°C to 1.0 at 250°C. Finally, the deposition rate was found to be 0.78 Å/cycle

As we have seen in all of the above discussion, the deposition of mixed oxides can be done in many cases, but achieving a film suitable for a specific application may be quite difficult. Specifically, there may be problems in achieving proper stochiometry or film morphology, among other film characteristics. In the remainder of this chapter, we will explore the concept of growing nanolaminate oxides byALD.

4.8.2 Nanolaminate Oxides

In the first part of this chapter, it has been demonstrated that there are many simple and mixed oxide thin films that can be deposited successfully by thermal ALD. Each has unique physical, chemical and electrical characteristics. For some applications, the ideal oxide doesn't exist, but it may be possible to prepare a multilayer film that would come close to this objective. Since this new technique is able to deposit very thin films (as thin as 10Å) in a very uniform and conformal way, such an approach may work in some cases. In other words, we can grow "designer" films, depending on the application.

As we may recall, some films grow as smooth amorphous layers and retain that morphology even under high temperature annealing. Other films may grow as rough films due to non-uniform nucleation. If the latter film has other characteristics that would be ideal for a certain application, it may be possible to combine the former film with it to grow a smooth multi-layered film with the novel characteristics. This would be referred to as a nanolaminate. Another example would be a thin film which has a high value of dielectric constant, but allows excessive current leakage. This film could be combined with another film which has minimal current leakage, but may have too low a dielectric constant. We would then have a nanolaminate film which has both a high dielectric constant, and that would also minimize leakage current.

In the following sections we will review a number of papers that address the question of creating nanolaminate films.

The as grown morphology of a thin film grown by thermal ALD may depend on the inherent characteristics of a specific film, on the surface upon which it is deposited and on the temperature at which it is deposited. Moreover, its morphology may change upon annealing. When a nanolaminate film is grown from two different films, the individual film morphologies may also depend on which two films are used to grow the nanolaminate. In this section, we will review this issue for nanolaminates formed from the simple oxides covered in earlier in this chapter.

Aluminum oxide is a simple oxide with many useful characteristics that make it an interesting film to use to form nanolaminates. As we have seen earlier in this Chapter, it can be grown as a smooth amorphous material which resists crystallization at high temperatures. Accordingly, it has been studied as one component of nanolaminates combined with many other materials, including tantalum [181, 185], titanium [182, 183, 200, 201], niobium [189, 196], zirconium [194, 196, 197, 198], hafnium [191, 192], silicon [195] and zinc [193].

Application of thin films of titanium dioxide to optical applications is attractive because of their usefulness in a variety of optical devices [182, 183]. However, as we have seen such films deposited by thermal ALD tend to grow with a rough surface. If a smoother film could be grown, the reduction of light absorption, and surface scattering, would enhance the application of such films. One example of a nanolaminar film that may meet the requirements for optical applications can be grown from alternate layers of titanium and aluminum oxides.

For example, in an early study [182] very thin alternating layers of aluminum oxide and titanium dioxide were deposited from TMA and $TiCl_4$ with hydrogen peroxide as the oxidizer, onto silicon, at 300°C Deposition was carried out in a vacuum chamber that was continuously pumped with a high capacity turbo molecular pump to 10^{-06} torr. The pure precursor vapors were introduced alternately at 1×10^{-04} torr for one second, with two second evacuations between the precursor pulses. A typical cycle timing diagram, is shown in Figure 4.35, with each cycle requiring six seconds.

Twenty pairs of 16Å thick films were grown, with the objective of creating a very smooth nanolaminar film. Because the growth rates of each layer was under 1Å/cycle, 20 cycles were needed to

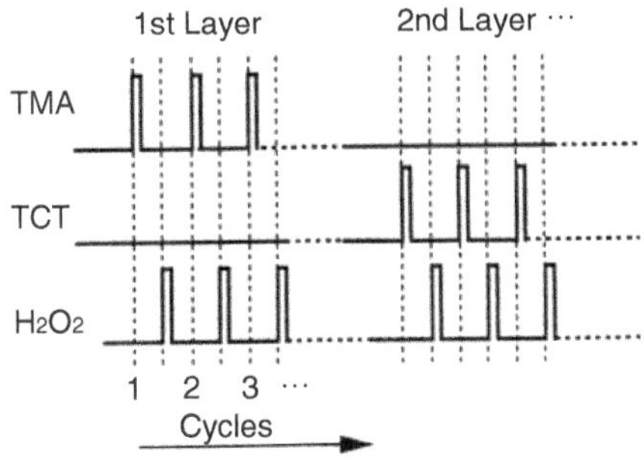

Figure 4.35 Timing cycles for growth of Al_2O_3/TiO_2 nanolaminate [182]. © *1994 The Japan Society of Applied Physics.*

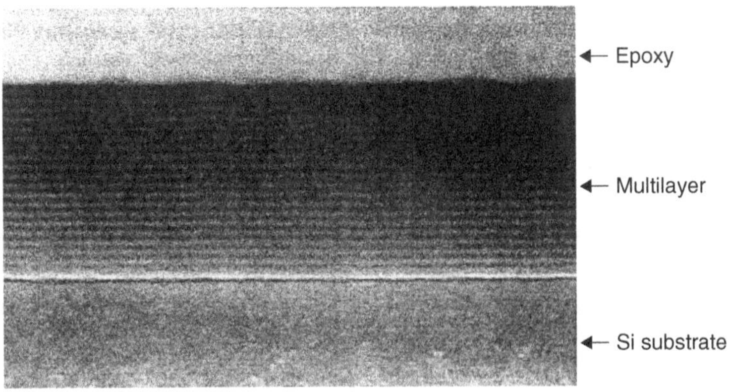

Figure 4.36 TiO_2/Al_2O_3 nanolaminate with 16Å thick layers [182]. © *1994 The Japan Society of Applied Physics.*

grow the Al_2O_3 layers, and 30 cycles were used for the TiO_2 layers. Figure 4.36 shows a cross-sectional TEM of the nanolaminate film.

The light lines are aluminum oxide, and the dark ones are the titanium oxide. The individual layers appear very uniform, and the complete nanolaminate is smooth. Clearly, the crystallization of the titanium oxide was suppressed by the use of the alternate aluminum oxide layers.

A similar study was done at about the same time [183], growing thin layers of titanium and aluminum oxides from titanium and aluminum halides ($TiCl_4$ and $AlCl_3$) using water as the oxidizer. However, in these experiments nitrogen was used as a purge gas so that films could be grown more quickly, with each cycle taking between one and two seconds. In this case the films of interest were considerably thicker than in the previous study, so that this higher deposition rate was very useful. Deposition was done on soda lime glass substrates at 500°C.

Nanolaminates were grown to an overall thickness of ~ 1 µm by growing 20 sets of binary layers (~ 500Å each). Each binary layer consisted of one Al_2O_3 and one TiO_2 layer, where the Al_2O_3 layer was held to a fraction of the binary layer thickness. The objective was to determine how thick a layer of Al_2O_3 was necessary to improve the film roughness found when pure TiO_2 was grown. It was found that Al_2O_3 layers that were 5–10% of the binary layer thickness produced nanolaminate films that were significantly smoother than pure TiO_2 films.

A more recent investigation explored the same nanolaminate deposited from TMA and $TiCl_4$, except that in this case water was used as the oxidant rather than hydrogen peroxide. Deposition was carried out, on bare silicon, at 200°C in a flow reactor with nitrogen as the purge gas [200]. Each binary layer was composed of 200Å thick films, and 5–10 such layers were formed to create the nanolaminate. Confirming the earlier result, the layers were very uniform in thickness with sharp edges between the Al_2O_3 and the TiO_2. Both layers were amorphous as deposited for the 5 laminate stack. However, when the 10 laminate stack was grown, there was one trace of crystallization in the innermost TiO_2 layer as can be seen in Figure 4.37.

Since these layers had been deposited onto silicon at 200°C, it is possible that this limited crystallization could be due to thermal stresses. When thinner layers are grown this effect was not seen [182].

As we have seen, nanolaminates composed of thin amorphous layers can be deposited on flat substrates. Next, it would be appropriate to evaluate how these nanolaminates behave when coating substrates with high aspect ratio holes or trenches. A nanolaminate composed of alternate layers of Al_2O_3 and Ta_2O_5 was grown from TMA and tantalum pentachloride at 200 - 300°C, using water as the oxidizer [191]. Conformal deposition over 12:1 aspect ratio trenches was achieved, as can be seen in Figure 4.38.

The nanolaminate stack consisted of alternate layers of five 14Å thick Al_2O_3 layers, and four 27Å thick Ta_2O_5 layers. As can be seen, the nanolaminate is extremely conformal, and all layers are amorphous and continuous.

Other issues as far as the morphology of these nanolaminates is concerned, include whether or not both of the alternating layers are always amorphous, do either crystallize and at what temperatures,

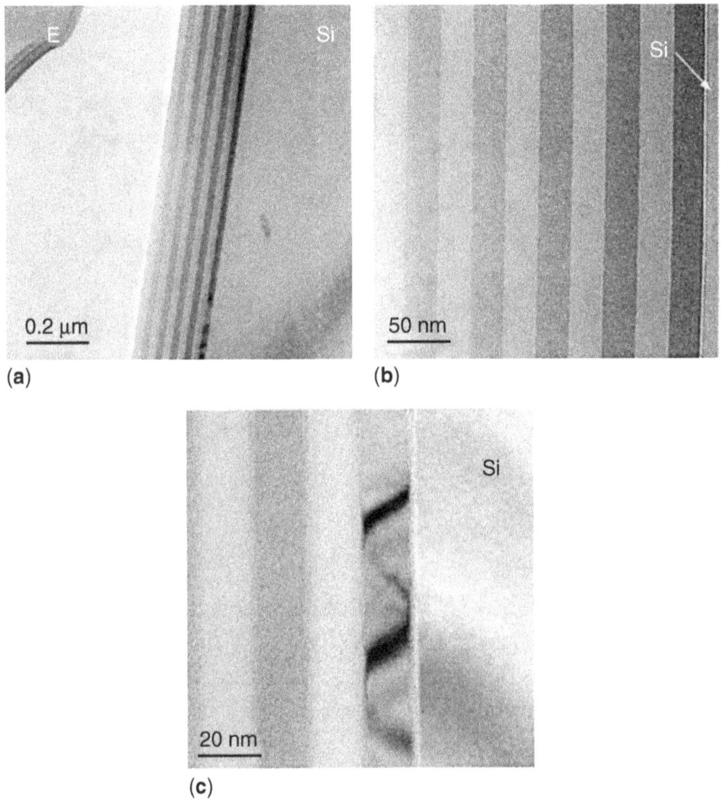

Figure 4.37 Nanolaminate of TiO_2 and Al_2O_3 – 10x stack. TEM image (c) shows detail of polycrystalline area of inner most TiO_2 layer of TEM image (a) [200]. *Reprinted from Appl. Surf. Sci. 243, D. R. G. Mitchell, D. J. Attard, K. S. Finnie, G. Triani, C. J. Barbe, C. Depagne , J. R. Bartlett, TEM and ellipsometry studies of nanolaminate oxide films prepared using atomic layer deposition, 585 © 2005 with permission from Elsevier.*

and finally can a smooth nanolaminate be grown even though one or both layers are crystallized. We have seen earlier, that thin amorphous Al_2O_3 layers can be effective in inhibiting the usual crystallization of a TiO_2 layer. It seems that the same is true for or a Ta_2O_5 layer.

Two other studies [198, 199] have examined the film morphology and smoothness of nanolaminates of HfO_2 and ZrO_2. Both of these films, on their own, tend to crystallize at the temperatures used for thermal ALD.

In the first investigation [199], thin films were grown on p-type <100> silicon wafers, with a 15Å thick native oxide layer. The

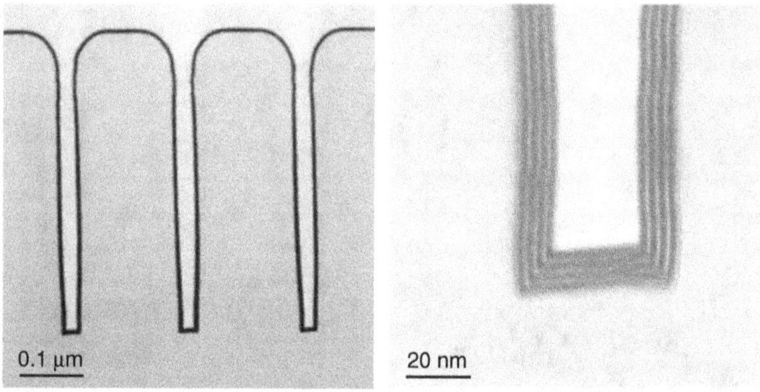

Figure 4.38 Al_2O_3/Ta_2O_5 nanolaminate coating 12:1 aspect ratio trenches [192]. *Reprinted from Thin Solid Films, 402, O. Sneh, R. B. Clark-Phelps, A. R. Londergan, J. Winkler, T. E. Seidel, Thin film atomic layer deposition equipment for semiconductor processing, 248 © 2002 with permission from Elsevier.*

precursors were $ZrCl_4$ and $HfCl_4$ with water as the oxidant, and depositions were carried out at 300°C. At this temperature a single ZrO_2 layer was fully crystalline, and a single HfO_2 layer was reported to be amorphous. After annealing at 500°C, each layer was crystalline.

Nanolaminates consisting of two and four layers were grown, with ZrO_2 and HfO_2 layers alternating. Individual layers were 45Å thick. Different results were obtained when the first layer was the Zr oxide, as compared to when the first layer was the Hf oxide. All of the layers were crystalline, even without annealing, in contrast to the results with individual layers. Also, those nanolaminates that began with a first layer of ZrO_2 had a rougher surface, as compared to those that started with a HfO_2 layer.

In a separate study [198], the same nanolaminate was examined with the objective of seeing whether or not crystallization may be prevented. In this case, deposition of the HfO_2 layer was from the tetrakis(dimethylamido) hafnium (IV), and for the ZrO_2 layer tetrakis(dimethylamido) zirconium (IV) was used. Water was used as the oxidant, and deposition was carried out at 200°C.

The total film thickness grown was on the order of 500Å. In one experiment the nanolaminate was composed of equal volumes of ZrO_2 and HfO_2. A second experiment created a similar nanolaminate with Al_2O_3 replacing HfO_2. As can be seen in Figure 4.39, the

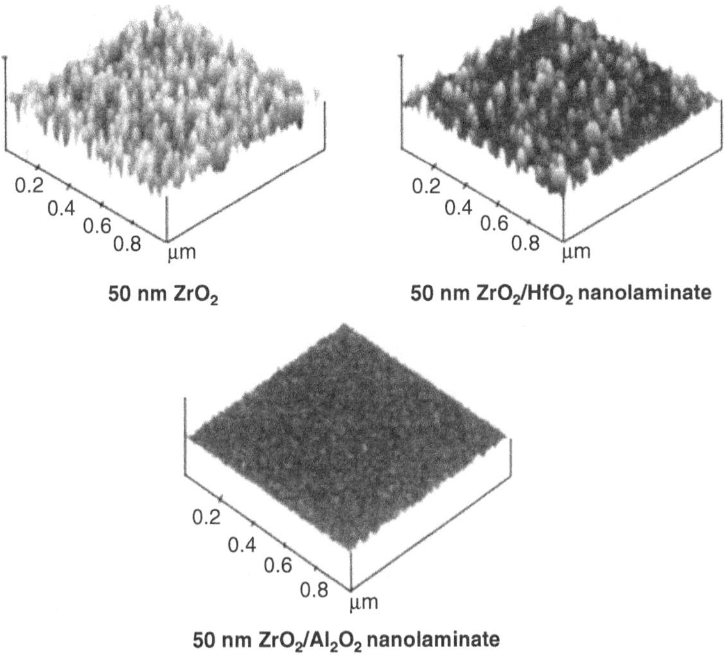

Figure 4.39 AFM surface plots of ZrO_2 thin film, and the same film modified with interspersed layers of HfO_2 or Al_2O_3. Height scale is 0 to 200Å [198]. *Reprinted from J. Cryst. Growth, 249, D. M. Hausmann and R. G. Gordon, Surface morphology and crystallinity control in the atomic layer deposition (ALD) of hafnium and zirconium oxide thin films, 251 © 2003 with permission from Elsevier.*

nanolaminate combining Hf and Zr oxides is smoother than a similar thickness layer of pure ZrO_2.

However, as we would have expected from the results described earlier in this chapter, the film with Al oxide replacing Hf oxide has a completely smooth surface. It is also noted that creating a nanolaminate using only 5Å of Al_2O_3 (approximately the thickness of one monolayer) interspersed with 100Å thick layers of ZrO_2, can still yield smooth films.

Finally, ZnO is a transparent semiconducting oxide with applications in flat panel displays and solar cells. As such, the fact that it grows as a crystalline film with increasing roughness as it gets thicker, can be a disadvantage in these applications. As we have already seen, one way to limit the roughness of these films is to grow a nanolaminate with Al_2O_3. In a recent study [193], it has been demonstrated that interrupting the growth of a ZnO layer with as little as one monolayer of Al_2O_3, can improve its surface roughness significantly. These results are consistent with the results just described for ZrO_2 films.

In the previous section, we examined the issues related to film morphology when growing nanolaminate films. In particular, we looked at whether one could create smooth surfaced films where rough surface simple oxide films would have been obtained. Also, whether or not a film was amorphous, polycrystalline or epitaxial when layered with another film was evaluated. Aside from these questions addressed by creating nanolaminates, there is the important issue regarding the electrical behavior of such films as dielectrics. In other words, there is considerable practical interest in creating thin dielectric films that have high dielectric constants, low leakage currents, and high breakdown strength. As we have seen in earlier in this chapter, in general, simple oxides have difficulty meeting all three conditions simultaneously.

One solution to this dilemma is the growth of mixed rather than simple oxides described earlier in this chapter. Another approach is to grow nanolaminates from two simple oxides. Several studies [181, 184, 185, 186, 189, 190] have looked at Ta_2O_5 laminated with HfO_2, ZrO_2, Al_2O_3, and Nb_2O_5. In addition to the Al_2O_3/Ta_2O_5 combination [181, 185], Al_2O_3 has also been laminated with Nb_2O_5, ZrO_2, and TiO_2 [189, 194, 197, 201]. Other studies have examined the electrical properties of ZrO_2/HfO_2 [186, 190] and $SrTiO_3/BaTiO_3$ [187, 188] nanolaminates. Although some of these investigations have been done to find a suitable replacement for a gate oxide in an integrated circuit, the solid state electronics of the interface between a nanolaminate and a bare silicon wafer is beyond the scope of this book. Therefore, these issues will not be considered.

One of the earliest studies of the electrical behavior of an oxide nanolaminate was for HfO_2/Ta_2O_5 [184]. Here an amorphous Ta_2O_5 film, which has a high dielectric constant and high breakdown strength, has an undesirably high leakage current at moderate applied voltages. Alternating layers of HfO_2, which can have low leakage current, can improve the overall electrical behavior of the nanolaminate.

The individual layers were deposited from $Ta(OC_2H_5)$, $HfCl_4$ and water at 325°C. The ALD reactor and deposition processes were the same as described earlier in this chapter for the

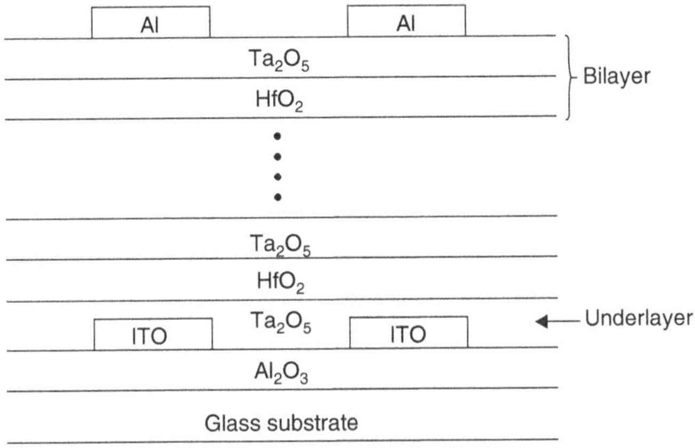

Figure 4.40 Nanolaminate structure of HfO_2/Ta_2O_5 film according to Kukli *et al* [184].

individual films. The total nanolaminate thickness was kept at 1700Å, Individual HfO_2/Ta_2O_5 bilayer thicknesses were fixed at 50, 100 or 200Å. The Ta_2O_5 layers ranged from 12 to 188Å, and the HfO_2 layer thicknesses were adjusted to maintain a constant thickness for all bilayers in a given nanolaminate. Since the nanolaminates were deposited onto glass, they were not annealed. The structure of the nanolaminates for these experiments is shown schematically in Figure 4.40, where the substrate is soda lime glass, and the Al_2O_3 layer serves as a barrier to diffusion of sodium into the nanolaminar film.

The study by Kukli *et al*. showed that the leakage current can be dramatically reduced by adjusting the layer thicknesses within each bilayer in the nanolaminate. In particular the leakage current for any of these nanolaminates is substantially less than either the pure HfO_2 film or the pure Ta_2O_5 film. The best nanolaminates are those with the greatest percentage of HfO_2, but these have the lowest values of permittivity. However, the film with 20% HfO_2 still has a respectable leakage current. In addition, this film has a permittivity value of ~ 25 which is essentially the same value as pure Ta_2O_5.

Clearly, these nanolaminates have far better electrical characteristics than the individual simple oxides themselves. The reason for this phenomenon is the morphology of the individual layers. By varying the individual thicknesses in a given bilayer, we can influence whether or not the separate films will be amorphous or polycrystalline. At the same time, the size of the individual crystals in a polycrystalline film can also be altered. Both of these effects will play a role in changing the leakage current and dielectric constant of these films.

Growth of HfO_2/Ta_2O_5 nanolaminates has also been carried out from $TaCl_5$ and $HfCl_4$ with water as the oxidant at 300°C [186, 190]. The primary purpose of creating the film, in this case, was to replace the native oxide in a silicon integrated circuit. As such, the overall film thickness was much thinner than in the previous studies ~ 100Å.

It consisted of a silicon nitride layer of 27.6Å with the remainder being the nanolaminate. The nanolaminate was composed of two bilayers, where each bilayer had 12Å of Ta_2O_5 and 22Å of HfO_2, and was annealed at 800°C for 60 sec. For this nanolaminate, a dielectric constant of 12 was measured, and the leakage current density was 2.0×10^{-7} A/cm^2. Because the dielectric constant of this 100 Å stack was much higher than pure SiO_2 (~ 3.9), it could be used as a replacement, or would be equivalent to, an SiO_2 film of 32.5Å thickness. In other words, the EOT (Equivalent Oxide Thickness) calculated from

$$\text{EOT} = T_{\text{stack}} \times (3.9/12) \text{ would be } 32.5 \text{ Å}.$$

Apparently, the presence of the underlying silicon nitride layer, between the bare silicon and the nanolaminate, was the cause of the observed low value for ε, as well as the very low leakage current. The same authors also grew ZrO_2/HfO_2 nanolaminates by the same technique with similar electrical results. However, they found that this nanolaminate [190] had an unexpectedly high breakdown strength of ~ 25 MV/cm.

All of the nanolaminates discussed, so far, involve either amorphous or polycrystalline layers. The experimental results indicate that the electrical characteristics of these multi layer stacks can be improved over the individual layers. Of course, the degree of improvement will depend on types of films chosen, as well as the film thicknesses and the deposition conditions. It is also possible to grow nanolaminates from single crystal (epi) films. These are referred to as superlattices, because the layers must adapt to the slightly different single crystal atomic spacing's. We will conclude this chapter with one example of such a superlattice nanolaminate.

Earlier in this chapter we noted that it is possible to grow the mixed oxides strontium and barium titanate with water as the oxidizer [160, 161]. The growth temperature was 325°C, to avoid decomposition of the precursers on the substrate surface, and produced amorphous films that were converted into polycrystalline form by annealing. In a separate paper [162], a different strontium precursor was used (dipivaloylmethane chelate of strontium [Sr(DPM)$_2$·2tetraene]), and epitaxial films were deposited onto epitaxial grade MgO substrates at 620°C. The oxidant in these experiments was N_2O. Very little information was reported on the self limiting nature of the process. In other words, was it a true ALD process? There was, however, a brief note where the authors said they were able to grow a "nearly monomolecular" layer per cycle for precursor flows of 3–9 sec duration.

Based on these results, these authors grew a strontium titanate/barium titanate nanolaminar film using the strontium precursor mentioned above, and the barium counterpart along with pure oxygen gas as the oxidizer [187, 188]. Deposition temperature was 620°C, and the substrate used was Nb doped (001)$SrTiO_3$.

The total stack thickness was 800 Å, with binary layers of 10, 20, and 40 Å thicknesses. Interestingly, the stack with the thinnest layers displayed an unusually high dielectric constant, as can be seen in Figure 4.41. In spite of these interesting results, two questions will have to be explored further. First, is the process truly self-limiting, and secondly how could one grow such a super lattice onto silicon substrates? Clearly, much additional research will be needed in order to understand this deposition technique.

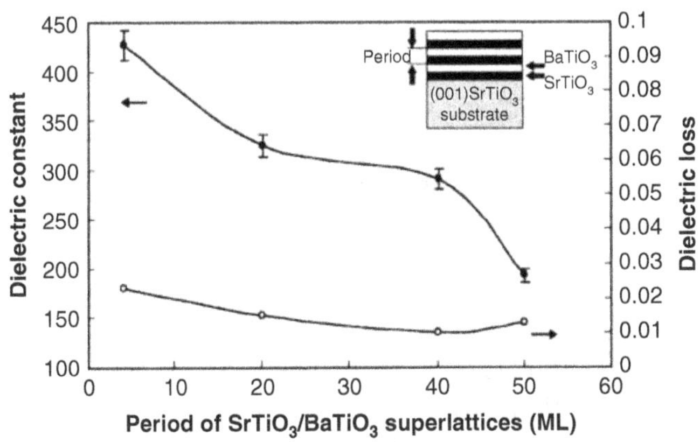

Figure 4.41 Dielectric properties of $SrTiO_3/BaTiO_3$ superlattice films [187]. © *1999 The Japan Society of Applied Physics.*

4.9 Multilayers

One issue that has been mentioned on several occasions, as we have reviewed different processes, relates to the interaction between the reactants and the substrate. Specifically, when the second reactant is a radical, it can attack the substrate in addition to promoting a reaction with the first reactant to form a monolayer of solid film.(in other words, activate the REALD process).

For example, when depositing an oxide onto an electrically conducting substrate (i.e. metals), using oxygen atoms as the free radical, these radicals are very likely to oxidize the underlying metal to some extent. On the other hand, if we deposit a conducting film onto an oxide substrate by using hydrogen atoms as the radical, there is a good possibility that the hydrogen will reduce the oxide to some extent. If either the oxide or the electrically conducting layer is very thin, a significant percentage of their thickness may be lost in the process.

One way to either eliminate or minimize such an effect would be to create an adsorbed multilayer composed of both reactants under conditions (i.e. low temperature) that allow them to remain adsorbed but unreacted. Then exposure of this multilayer to a flux of energetic but inert radicals could provide the necessary energy to convert this multilayer to the thin film desired. Repeating this process would enable the growth of films of the desired thickness [202].

As one example of such a process, TiN was grown from $TiCl_4$ and NH_3 using nitrogen atoms generated in a remote plasma source. Deposition temperature was 200°C. The radicals were generated inductively with a coil operated at 350 watts and at a frequency of 13.56 MHz. A total of 400 deposition cycles was carried out.

The resulting film was the characteristic color of TiN, and was electrically conductive. It was also hard and adherent. It could not be scratched, and easily passed the scotch tape test. These preliminary experiments confirmed that a multilayer had indeed been formed that could be converted to a solid film using a radical exposure.

If an inert noble gas had been used to generate highly energetic electronically excited atoms rather than a reactive radical (i.e. nitrogen atoms), then other processes could be developed. For example, exposing a substrate to TMA and water should form a multilayer, and the energy released from the de-excitation of a noble gas metastable atom should produce a layer of aluminum oxide. Such a process could be particularly useful if thin layers of aluminum oxide needed to be deposited onto plastic substrates, since the carbon in the plastic hydrocarbon could otherwise be attacked.

References

1. V. Miikkulainen, M. Leskelä, M. Ritala, R. L. Puurunen, *Appl. Phys. Rev*, 113, 021301 (2013).
2. H. B. Profijt, S. E. Potts, M. C. M. van de Sanden, W. M. M. Kessels, *J. Vac. Sci. Technol. A* 29(5), 050801 (2011).
3. G. Oya, M. Yoshida and Y. Sawada, *Appl. Phys. Lett.* 51, 1143 (1987).
4. G.S. Higashi and C.G. Fleming, *Appl. Phys. Lett.* 55, 1963 (1989).
5. J.F. Fan, K. Sugioka and K. Toyoda, *Mat. Res. Soc. Symp. Proc.* 222, 327 (1991).
6. J-F. Fan, K. Sugioka and K. Toyoda, *Jpn. J. Appl. Phys.* 30, L1139 (1991).
7. H. Kumagai, K. Toyoda, M. Matsumoto and M. Obara, *Jpn. J. Appl. Phys.* 32, 6137 (1993).
8. E-L. Lakomaa, A Root and T. Suntola, *Appl. Surf. Sci.* 107, 107 (1996).
9. S.J. Yun, K-H. Lee, J. Skarp, H-R. Kim and K-S. Nam, *J. Vac. Sci. Technol. A*15, 2993 (1997).
10. K. Kukli, M. Ritala and M. Leskelä, *J. Vac. Sci. Technol. A* 15, 2214 (1997).
11. L. Hiltunen, H. Kattelus, M. Leskelä, M. Makelä, L. Ninistö, E. Nykänen, P. Soininen and M. Tiitta, *Mat. Chem Phys.* 28, 379 (1991).
12. M. Juppo, A. Rahtu, M. Ritala, and M. Leskala, *Langmuir* 16, 4034 (2000).
13. R. Matero, A. Rahtu, M. Ritala, M. Leskala, and T. Sajavaara, *Thin Solid Films* 368, 1 (2000).
14. M. D. Groner, J. W. Elam, and S. M. George, AVS 2nd International Conference on Microelectronics and Interfaces, Feb. 2001, Santa Clara, CA.
15. R. L. Puurunen, *Appl. Surf. Sci.*245 (2005).
16. J. B. Kim, D. R. Kwon, K. Chakrabarti, C. Lee, K. Y. Oh, and J. H. Lee, *J. Appl. Phys.* 92, 6739 (2002).
17. J. Kim, K. Chakrabarti, J. Lee, K. Y. Oh, and C. Lee, *Mater. Chem. Phys.* 78, 733 (2003).
18. S. Jakschik, U. Schroeder, T. Hecht, G. Dollinger, A. Bergmaier, and J. W. Bartha, *Mater. Sci. Engr.* B107, 251 (2004).
19. S. C. Ha, E. Choi, S. H. Kim, and J. S. Roh, *Thin Solid Films* 476, 252 (2005).
20. Y-S. Min, Y. J. Cho and C. S. Hwang, *Chem. Mater.* 17, 626 (2005).
21. W, S. Jeon, S. Yang, C. S. Lee, and S. W. Kang, *J. Electrochem. Soc.* 149, C306 (2002).

22. R. L. Puurunen, *Chem. Vap. Deposition* 9, 249 (2003).
23. R. L. Puurunen, *Chem. Vap. Deposition* 9, 327 (2003).
24. R. L. Puurunen, *J. Appl. Phys.* 97, 121301 (2005).
25. A. W. Ott, J. W. Klaus, J. M. Johnson, S. M. George, *Thin Solid Films* 292 (1997).
26. R. Kuse, M. Kundu, T. Yasuda, N. Miyata, and A. Toriumi, *J. Appl. Phys.* 94, 10 (2003).
27. M. D. Groner, F. H. Fabreguette, J. W. Elam, and S. M. George, *Chem. Mater* 16 (2004).
28. C. A. Wilson, R. K. Grubbs, and S. M. George, *Chem. Mater.* 17 (2005).
29. K. Grigoras, S. Franssila, and V.-M. Airaksinen, *Thin Solid Films* 516, 5551 (2008).
30. S. D. Elliot, G. Scarel, C. Wiemer, M. Fanciulli, G. Pavia, *Chem. Mater.*18 (2006).
31. P. I. Räisänen, M. Ritala, M. Leskelä, *J. Mater. Chem.* 12, 1415 (2002).
32. M. Ritala, K. Kukli, A. Rahtu, P. I. Räisänen, M. Leskelä, T. Sajavaara, and J. Keinonen, *Science* 288, 319 (2000).
33. A. W. Laubengayer and W. F. Gilliam, *J. Am. Chem. Soc.* 63, 477 (1941).
34. J. W. Lim and S. J. Yun, Jpn. *J. Appl. Phys.* 42, L 663 (2003).
35. K. Y. Park, H. I. Cho, J. H. Lee, S. B. Bae, C. M. Jeon, J. L. Lee, D. Y. Kim, C. S. Lee and J. H. Lee, *Phys. Stat. Sol.* (c) 0, 2351 (2003).
36. S. J. Yun, Y. W. Ko and J. W. Lim, *Appl. Phys. Lett.* 85, 4896 (2004).
37. S. J. Yun, J. W. Lim and J. H. Lee, *Electrochem. Solid-State Lett.* 7, C-13 (2004).
38. A. Sherman, *Chemical Vapor Deposition*, Noyes Publications, New Jersey, 1987.
39. E. Langereis, M. Creatore, S. B. S. Heil, M. C. M. Van de Sanden, W. M. M. Kessels, *Appl. Phys. Lett.* 89, 081915 (2006).
40. E. Langereis, M. Bouman, J. Keijmel, S. B. S. Heil, M. C. M. van de Sanden, W. M. M. Kessels, *ECS Trans.* 16(4), 247 (2008).
41. W. S. Kim, D. Y. Moon, B. W. Kang, J. W. Park, J. G. Park, *J. Korean Phys. Soc.* 55, 55 (2009).
42. A. Fujishima, K. Honda, *Nature* 238, 38 (1972).
43. A.L. Linsebigler, G. Lu, J.T. Yates, *Chem. Rev.* 95, 735 (1995).
44. R.F. Howe, M. Grätzel, *J. Phys. Chem.* 91, 3906 (1987).
45. M.R. Hoffmann, S.T. Martin, W. Choi, D.W. Bahnemann, *Chem. Rev.* 95, 69 (1995).
46. D.W. Bahnemann, M. Hilgendorff, R. Memming, J. *Phys. Chem.* B 101, 4265 (1997) .
47. C. Di Valentin, G. Pacchioni, A. Selloni, *Phys. Rev. Lett.* 97, 166803 (2006).
48. M.-L. Kääriäinen, T.O. Kääriäinen, D.C. Cameron, *Thin Solid Films* 517, 6666 (2009).
49. R.Wang, K. Hashimoto, A. Fujishima, M. Chikuni, E.Kojima,A. Kitamura, M. Shimohigoshi, T. Watanabe, *Nature* 388, 431 (1997).

50. Akira Fujishima *, Xintong Zhang, C. R. *Chimie* 9, 750 (2006).
51. M. Ritala, M. Leskelä, E. Nykänen, P. Soininen and L. Niinistö, *Thin Solid Films* 225, 288 (1993).
52. J. Aarik, A. Aidla, T.Uustare, V. Sammelselg, J. *Crystal Growth* 148, 268 (1995).
53. J. Aarik, A. Aidla, H. Mändar, T. Uustare, M. Schuisky and A. Hårsta, *J.Cryst. Growth* 242, 189 (2002).
54. K.S. Finnie, G. Triani, K.T. Short, D.R.G. Mitchell, D.J. Attard, J.R. Bartlett, C.J. Barbé, *Thin Solid Films* 440, 109 (2003).
55. D.R.G. Mitchell, D.J. Attard, G. Triani, *Thin Solid Films*, 441, 85 (2003).
56. R. Matero, A. Rahtu, and M. Ritala, *Chem. Mater.* 13, 4506 (2001).
57. M.-L. Kääriäinen, D.C. Cameron, *Surface Science* 606, L22 (2012).
58. V. Pore, M. Heikkilä, M. Ritala, M. Leskelä, S. Areva, *J. Photochem. Photobiol. A* 177, 68 (2006).
59. M.-L. Kääriäinen, D.C. Cameron, *Thin Solid Films* 526, 212 (2012).
60. N. G. Kubala, P. C. Rowlette, C. A. Wolden, *J. Phys. Chem.* C 113, 16307 (2009).
61. N. G. Kubala, C. A. Wolden, *Thin Solid Films* 518, 6733 (2010).
62. K. Kukli, M. Ritala, M. Schuisky, M. Leskelä, T. Sajavaara, J. Keinonen, T.Uustare , A. Hårsta, *Chem. Vap. Deposition* 6, 303 (2000).
63. K. Kukli, A. Aidla, J. Aarik, M. Schuisky, A. Hårsta, M. Ritala and M. Leskelä, *Langmuir* 16, 8122 (2000).
64. J. Aarik, A. Aidla, T. Uustare, K. Kukli, V. Sammelselg, M. Ritala and M.Leskelä, *Appl. Surf. Sci.* 193, 277 (2002).
65. M. Schuisky, J. Aarik, K. Kukli, A. Aidla and A. Hårsta, *Langmuir* 17, 5508 (2001).
66. V. Pore, T. Kivelä, M. Ritala, M. Leskelä, *Dalton Trans.* 6467 (2008).
67. R.J. Narayan, N.A. Monteiro-Riviere, R.L. Brigmon, M.J. Pellin, J.W. Elam, *JOM* 61, 12 (2009).
68. M. Ritala, M. Leskelä, L. Niinistö, P. Haussalo, *Chem. Mater.* 5, 1174 (1993).
69. A. Rahtu, M. Ritala, *Chem. Vap. Deposition* 8, 21 (2002).
70. J. Aarik, A. Aidla, T. Uustare, M. Ritala, M. Leskelä, *J. Appl. Sci.* 161, 385 (2000).
71. V.R. Rai and S. Agarwal, *J. Phys. Chem. C Lett.* 112, 9552 (2008).
72. A. Niskanen, K. Arstila, M. Leskelä, M. Ritala, *Chem. Vap. Deposition* 13, 152 (2007).
73. M. Ritala, M. Leskelä, E. Rauhala, *Chem. Mater.* 6, 556 (1994).
74. J. Aarik, J. Karlis, H. Mändar, T. Uustare, V. Sammelselg, *Appl. Surf. Sci.* 181, 339 (2001).
75. V. Pore, A. Rahtu, M. Leskelä, M. Ritala, T. Sajavaara, J. Keinonen, *Chem. Vap. Deposition* 10, 143 (2004).
76. I. Jõgi, M. Pärs, J. Aarik, A. Aidla, Matti Laan, J. Sundqvist, L. Oberbeck, J. Heitmann, K. Kukli, *Thin Solid Films* 516, 4855 (2008).

77. A.P. Alekhin, S.A. Gudkova, A.M. Markeev, A.S. Mitiaev, A.A. Sigarev, V.F. Toknova, *Appl. Surf. Sci.* 257, 186 (2010).
78. H. Y. Jeong, Y. I. Kim, J. Y. Lee, S. Y. Choi, Nanotechnology 21, 115203 (2010).
79. Q. Xie, Y.-L. Jiang, C. Detavernier, D. Deduytsche, and R. L. Van Meirhaeghe, G.-P. Ru, B.-Z. Li, X.-P. Qu, *J. Appl. Phys.* 102, 083521 (2007).
80. Q. Xie, J. Musschoot, D. Deduytsche, R. L. Van Meirhaeghe, C. Detavernier, S.Van den Berghe, Y.-L. Jiang, G.-P. Ru, B.-Z. Li and X.-P. Qu, *J. Electrochem.Soc.* 155, H688(2008).
81. C.-S. Lee , J. Kim, J.Y. Son, W. Choi, H. Kim, Appl. Catal. B 91, 628 (2009).
82. C.-S. Lee, J. Kim, G.H. Gu, D.-H. Jo, C.G. Park, W. Choi, H. Kim, *Thin Solid Films* 518, 4757 (2010).
83. T.O.Kääriäinen, D.C. Cameron, M. Tanttari, Plasma Process. Polym. 6, 631 (2009).
84. T.O. Kääriäinen, S. Lehti, M.-L. Kääriäinen, D.C. Cameron, *Surf. Coat. Tech.* 205, S475 (2011).
85. A. Sarkar, S.E. Pitts, S.A. Rushworth, F. Roozeboom, M.C.M. van den Sanden, W.M.M. Kessels, *ECS Transact.* 33, 385 (2010).
86. J. Niinistö, T. Hatanpää, M. Kariniemi, M. Mäntymäki, L. Costelle, K. Mizohata, K. Kukli, M. Ritala, M. Leskelä, *Chem. Mat.* 24, 2002 (2012).
87. I. Jõgi, J. Aarik, M. Laan, J. Lu, K. Kukli, H. Käämbre, T. Sajavaara, T. Uustare, *Thin Solid Films* 510, 39 (2006).
88. I. Jõgi, K. Kukli, J. Aarik, A. Aidla, J. Lu, Mat. Sci. Semicon. Proc. 9, 1084 (2006).
89. D.M. King, X. Liang, C.S. Carney, L.F. Hakim, P. Li, and A.W. Weimer, *Adv. Funct. Mater.* 18, 607 (2008).
90. J. Lim, C. Lee, *J. Alloys Compd.* 449, 371 (2008).
91. S.K. Kim, C.S. Hwang, S.-H. K. Park, S.J.Yun, *Thin Solid Films* 478, 103 (2005).
92. S. Muthukumar, N.W. Emanetoglu, G. Patounakis, C.R. Gorla, S.Liang, Y. Lu, *J. Vac. Sci. Technol.* A 19, 1850 (2001).
93. M. Schuisky, J. W. Elam, S. M. George, *Appl. Phys. Lett.* 81, 180 (2002).
94. S.J. Lim, S. Kwon, H. Kim, *Thin Solid Films* 516, 1523 (2008).
95. D. D. Fong, J. A. Eastman, S. K. Kim, T. T. Fister, M. J. Highland, P. M. Baldo, P. H. Fuoss, *Appl. Phys. Lett.* 97, 191904 (2010).
96. J. S. Na, Q. Peng , G. Scarel , G. N. Parsons, *Chem. Mater.* 21, 5585 (2009).
97. D.M. King, X. Liang, C.S. Carney, L.F. Hakim, P. Li, and A.W. Weimer, *Adv. Funct. Mater.* 18, 607 (2008).
98. J. Jur , W.J. Sweet III , C. J. Oldham , G.N. Parsons, *Adv. Funct. Mater.* 21, 1993 (2011).

99. D.A. Mourey, D.A. Zhao, J. Sun, T.N. Jackson, *IEEE Trans. Electron Devices* 57, 530 (2010).
100. K.-S. An, W. Cho, B.K. Lee, S.S. Lee, C.G. Kim, *J Nanosci. Nanotechno.* 8, 4856 (2008).
101. M. Ritala and M. Leskelä, *Appl. Surf. Sci.* 75, 333 (1994).
102. M. Copel, M. Gribelyuk, and E. Gusev, *Appl. Phys. Lett.* 76, 436 (2000).
103. M. Cassir, F. Goubin, C. Bernay, P. Vernoux, and D. Lincot, *Appl. Surf. Sci.* 193, 120 (2002).
104. J. Aarik, A. Aidla, H. Mandar, T. Uustare, and V. Sammelselg, *Thin Solid Films* 408, 97 (2002).
105. K. Kukli, M. Ritala, J. Aarik, T. Uustare, and M. Leskala, *J. Appl. Phys.* 92, 1833 (2002).
106. S.K. Dey, C.-G. Wang, D. Tang, M.J. Kim, R.W. Carpenter, C. Werkhoven, and E. Spiro, *J. Appl. Phys.* 93, 4144 (2003).
107. K. Kukli, K. Forsgren, J. Aarik, T. Uustare, A. Aidla, A. Niskanen, M. Ritala, M. Leskala, and A. Harsta, *J. Cryst. Growth* 231, 262 (2001).
108. K. Kukli, M. Ritala, T. Uustare, J. Aarik, K. Forsgren, T. Sajavaara, M. Leskala, and A. Harsta, *Thin Solid Films* 410, 53 (2002).
109. K. Kukli, M. Ritala, and M. Leskala, *Chem. Vap. Deposition* 6, 297 (2000).
110. R. Matero, M. Ritala, M. Leskelä, A. C. Jones, P. A. Williams, J. F. Bickley, A. Steiner, T. J. Leedham, and H. O. Davies, *J. Non-Cryst. Solids* 303, 24 (2002).
111. Y. Kim, J. Koo, J. Han, S. Choi, H. Jeon and C. G. Park, *J. Appl. Phys.* 92, 5443 (2002).
112. J. Y. Kim, S. H. Kim, H. Seo, J. H. Kim and H. Jeon, *Electrochem. Solid-State Lett.* 8, G82 (2005).
113. K. Endo and T. Tatsumi, *Mat. Res. Soc. Symp. Proc.* Vol. 745, N.2.8.1 (2003).
114. K. Endo and T. Tatsumi, *Jpn. J. Appl. Phys.* 42, L685 (2003)
115. Raija Matero, M. Ritala, M. Leskelä, T. Sajavaara, A.C. Jones, J.L. Roberts, *Chem. Mate*r. 16, 5630 (2004).
116. D. M. Hausmann, E. Kim, J. Becker, and R. G. Gordon, *Chem. Mater.* 14, 4350 (2002).
117. Päiväsaari, J., Dezelah, C.L., Back, D., El-Kaderi, H.M., Heeg, M.J., Putkonen, M.,Niinistö, L., and Winter, C.H., *J. Mater. Chem.* 15, 4224 (2005).
118. J. Niinistö, M. Putkonen, L. Niinistö, K. Kukli, M. Ritala, M. Leskelä, *J. Appl. Phys.*, 95, 84 (2004).
119. C.L. Dezelah, Jaakko Niinistö, K. Kukli, F. Munnik, J. Lu, M. Ritala, M. Leskelä, L. Niinistö, *Chem. Vap. Deposition* 14, 358 (2008).
120. M. Ritala, M. Leskelä, L. Niinistö, T. Prohaska, G. Friedbacher and M. Grasserbauer, *Thin Solid Films* 250, 72 (1994).

121. J. Aarik, A. Aidla, A.-A. Kiistler, T. Uustare, and V. Sammelselg, *Thin Solid Films* 340, 110 (1999).
122. E. P. Gusev, E. Cartier, D. A. Buchanen, M. A. Gribelyuk, M. Copel, H. Okorn- Schmidt, and C. D'Emic, *Microelectronic Engr.* 59, 341 (2001).
123. M.-Y. Ho, H. Gong, G. D. Wilk, B. W. Busch, M. L. Green, P. M. Voyles, D. A. Muller, M. Bude, W. H. Lin, A. See, M. E. Loomans, S. K. Lahiri, and P. I. Raisanen, *J. Appl. Phys.* 93, 1477 (2003).
124. D. H. Triyoso, R. I. Hegde, J. Grant, P. Fejes, R. Liu, D. Roan, M. Rowan, D. Werho, R. Rai, L. B. La, J. Baker, C. Garza, T. Guenther, B. E. White, Jr., and P. J. Tobin, *J. Vac. Sci. Technol. B* 22, 2121 (2004).
125. H. B. Park, M. Cho, J. Park, S. W. Lee, C. S. Hwang, and J. Jeong, *Electrochem. and Solid State Lett.* 7, F25 (2004).
126. K. Kukli, M. Ritala, T. Sajavaara, J. Keinonen, and M. Leskala, *Thin Solid Films* 416, 72 (2002).
127. K. Forsgren, A. Harsta, J. Aarik, A. Aidla, J. Westlinder, and J. Olsson, *J. Electrochem. Soc.* 149, F139 (2002).
128. J. Aarik, A. Aidla, H. Mandar, V. Sammelselg, and T. Uustare, *J. Crystal Growth* 220, 105 (2000).
129. J. Aarik, J. Sundqvist, A. Aidla, J. Lu, T. Sajavaara, K. Kukli, , and A. Harsta,, *Thin Solid Films* 418, 69 (2002).
130. J. Sundqvist, A. Harsta, J. Aarik, K. Kukli, A. Aidla, *Thin Solid Films* 427, 147 (2003).
131. J. F. Conley, Jr., Y. Ono, W. Zhuang, D. J. Tweet, W. Gao, S. K. Mohammed, and R. Solanki, *Electrochem. Solid-State Lett.* 5, C57 (2002).
132. J. F. Conley, Jr., Y. Ono, D. J. Tweet, W. Zhuang, and R. Solanki, *J. Appl. Phys.* 93, 712 (2003).
133. J. F. Conley, Jr., Y. Ono, R. Solanki, G. Stecker, and W. Zhuang, *Appl. Phys. Lett.* 82, 3508 (2003).
134. D. M. Hausmann, E. Kim, J. Becker, and R. G. Gordon, *Chem. Mater.* 14, 4350 (2002).
135. M. Suzuki, T. Magara, T. Takahashi, and H. Shinriki, AVS ALD 2002 Conference, Seoul, Korea.
136. K. Kukli, M. Ritala, T. Sajavaara, J. Keinonen, and M. Leskelä, *Chem. Vap. Deposition* 8, 199 (2002).
137. K. Kukli, M. Ritala, M. Leskelä, T. Sajavaara, J. Keinonen, A.C. Jones, J.L. Roberts, *Chem. Mater.* 15, 1722 (2003).
138. W. Cho, K.-S. An, T.-M. Chung, C. G. Kim, B.-S. So, Y.-H. You, J.-H. Hwang, D. Jung, Y. Kim, *Chem. Vap. Deposition* 12, 665 (2006).
139. R. O'Kane, J. Gaskell, A. C. Jones, P.R. Chalker, K. Black, M. Werner, P. Taechakumput, S. Taylor, P.N. Heys, R. Odedra, *Chem. Vap. Deposition* 13, 609 (2007).
140. Minha Seo , Seong Keun Kim , Yo-Sep Min and Cheol Seong Hwang, *J. Mater. Chem.* 21, 18497 (2011).

141. A. L. Dawar and J. C. Joshi, *J. Mater. Sci.* 19, 1 (1984).
142. A. Rosental, A. Tarre, A. Gerst, J. Sundqvist, A. Harsta, A. Aidla, J, Aarik, V. Sammelselg and T. Uustare, *Sensors Actuators B* 93, 552 (2003).
143. M. Utriainen, H. Lattu, H. Viirola, L. Niinistö, Roland Resch, G. Friedbacher, *Mikrochim. Acta* 133, 119 (2000).
144. H. Viirola and L. Niinisto, *Thin Solid Films* 249, 144 (1994).
145. C. Ducso, N. Q. Khanh, Z. Horvath, I. Barsony, M. Utriainen, S. Lehto, M. Nieminen and L. Niinisto, *J. Electrochem. Soc.* 143, 683 (1996).
146. G. Natarajan, D.C. Cameron, *Appl Phys A* 95, 631 (2009).
147. A.J. Niskanen, Aapo Varpula, M. Utriainen. G. Natarajan2, D.C. Cameron, S.Novikov, V.-M. Airaksinend, J. Sinkkonen, S. Franssila, *Sensors and Actuators B* 148, 227 (2010).
148. T. Asikainen, M. Ritala and M. Leskelä, *J. Electrochem. Soc.* 141, 3210 (1994).
149. T. Asikainen, M. Ritala and M. Leskelä, *J. Electrochem. Soc.* 142, 3538 (1995).
150. M. Pessa, R. Mäkelä and T. Suntola, *Appl. Phys. Lett.* 38, 131 (1981).
151. K. Kukli, M. Ritala, R. Matero, M. Leskelä, *J. Crystal Growth* 212, 459 (2000).
152. K. Kukli, M. Ritala and M. Leskelä, *J. Electrochem. Soc.*, 142 1670 (1995).
153. K. Kukli, J. Aarik, A. Aidla, H. Siimon, M. Ritala and M. Leskelä, *Appl. Surf. Sci.* 112, 236 (1997).
154. H-J. Song, W. Koh, and S-W. Kang, *Mater. Res. Soc. Symp. Proc.* 567, 469 (1999).
155. H. J. Song, C. S. Lee and S. W. Kang, *Electrochem. Solid-State Lett.* 4, F13 (2001).
156. A. Niskanen, U. Kreissig, M. Leskelä, M. Ritala, *Chem. Mater.* 19, 2316 (2007).
157. K. Kukli, J. Aarik, A. Aidla, K. Forsgren, J. Sundqvist, A. Harsta, T. Uustare, H. Mandar, and A-A. Kiisler, *Chem. Mater.* 13, 122 (2001).
158. J. Sundqvist, H. Hogberg, and A. Harsta, *Chem. Vap. Deposition* 9, 245 (2003).
159. C.L. Dezelah, M.K. Wiedmann, K. Mizohata, R. J. Baird, L. Niinistö, C. H. Winter, *J. am. Chem. Soc* 129, 12370 (2007).
160. M. Vehkamäki, T. Hänninen, M. Ritala, M. Leskelä, T. Sajavaara, E. Rauhala, and J. Keinonen, *Chem. Vap. Deposition* 7, 75 (2001).
161. M. Vehkamaki, T. Hatanpaa, T. Hanninen, M. Ritala, and M. Leskela, *Electrochem. and Solid State Ltrs.* 2, 504 (1999)
162. Z. Wang and S. Oda, Jpn. *J. Appl. Phys.* 37, 942 (1998).
163. M. Vehkamäki, M. Ritala, M. Leskelä, A. C. Jones, H. O. Davies, T. Sajavaara, and E. Rauhala, *J. Electrochem. Soc.* 151, F69 (2004).
164. M. Nieminen, T. Sajavaara, E. Rauhala, M. Putkonen, and L. Niinistö, *J. Mater. Chem.* 11, 2340 (2001).

165. D. H. Triyoso, H. Li, R. I. Hegde, Z. Yu, K. Moore, J. Grant, B. E. White, Jr. and P. J. Tobin, *J. Vac. Sci. Technol.* B23, 2480 (2005).
166. T. Kawahara, K. Torii, S. Fukuda, T. Maeda, A. Horiuchi, H. Ito, A. Muto, Y. Kato, and H. Kitajima, *Mater. Res. Soc. Symp. Proc* 745, N5.6.1 (2003).
167. B. Crivelli, M. Alessandri, S. Alberici, F. Cazzaniga, D. Dekadjevi, J. W. Maes, G. Ottaviani, G. Pavia, G. Queirolo, S. Santucci, with F. Zanderigo, *Mater. Res. Soc. Symp. Proc* 745, N5.3.1 (2003).
168. B. Crivelli, M. Alessandri, S. Alberici, D. Brazzelli, A.C. Elbas, S. Frabboni, G. Ghidini, J.W. Maes, G. Ottaviani, G. Pavia, and C. Wiemer, Presented at Spring 2003 MRS Symposium, San Francisco, CA.
169. R. G. Gordon, J. Becker, and S. Suh, *Chem. Mater.* 13, 2463 (2001).
170. Y. Senzaki, S. Park, H. Chatham, and L. Bartholomew, *J. Vac. Sci. Technol.* A22, 1175 (2004).
171. Y. Okuyama, C. Barelli, C. Tousseau, S. Park, and Y. Senzaki, J. *Vac. Sci. Technol. A*23, 1 (2005).
172. W-H. Nam and S-W Rhee, *Electrochem. and Solid State Ltrs.*7, C55 (2004).
173. S-W. Kang, S-W. Rhee, and S. M. George, *J. Vac. Sci. Technol. A*22, 2392 (2004).
174. W-K. Kim, S-W. Kang, N-I Lee, J-H. Lee, and H-K. Kang, *J. Vac. Sci. Technol. A2O*, 2096 (2002).
175. W-K. Kim, S-W. Kang, and S-W. Rhee, *J. Vac. Sci. Technol.* A21, L16 (2003).
176. Z. M. Rittersma, E. Naburgh, T. Dao, A. H. C. Hendricks, W. F. A. Besling, E. Tois, E. Vainonen-Ahlgren, M. Tuominen, and S. Haukka, *Electrochem. and Solid State Ltrs.* 6, F21 (2003).
177. J. Harjuoja, T. Hatanpaa, M. Vehkamäki, S. Väyrynen, M. Putkonen, L. Niinistö, M. Ritala, M. Leskelä and E. Rauhala, *Chem. Vapor Dep.* 11, 362 (2005).
178. W. J. Lee, I. K. You, S. O. Ryu, B. G. Yu, K. I. Cho, S. G. Yoon and C. S. Lee, *Jpn. J. Appl. Phys.* 40, 6941 (2001).
179. B. G. Chae, W. J. Lee, I. K. You, S. O. Ryu, M. Y. Jung and B. G. Yu, *Jpn. J. Appl. Phys.* 41, L729 (2002).
180. D. S. Kil, J. M. Lee and J. S. Roh, *Chem. Vap. Deposition* 8, 195 (2002).
181. H. Kattelus, M. Ylilammi, J. Saarilahti, J. Anston and S. Lindfors, *Thin Solid Films* 225, 296 (1993).
182. H. Kumagai, M. Matsumoto, Y. Kawamura, K. Toyoda and M. Obara, *Jpn. J. Appl. Phys.* 33, 7086 (1994).
183. M. Ritala, M. Leskala and N. Niinisto, *Thin Solid Films* 249, 155 (1994).
184. K. Kukli, J. Ihanus, M. Ritala and M. Leskala, *Appl. Phys. Lett.* 68, 3737 (1996).
185. K. Kukli, J. Ihanus, M. Ritala and M. Leskala, *J. Electrochem. Soc.* 144, 300 (1997).
186. H. Zhang, R. Solanki, B. Roberds, G. Bai and I. Banerjee, *J. Appl. Phys.* 87, 1921 (2000).

187. Z. Wang, T. Yasuda, S. Hatatani and S. Oda, Jpn. J. *Appl. Phys.* 38, 6817 (1999).
188. Z. Wang and S. Oda, *J. Electrochem. Soc.* 147, 4615 (2000).
189. K. Kukli, M. Ritala and M. Leskala, *J. Electrochem. Soc.* 148, F35 (2001).
190. H. Zhang, R. Solanki, *J. Electrochem. Soc.* 148, F63 (2001).
191. R. B. Clark-Phelps, A. Srivastava, L. Cleveland, T. Seidel and O. Sneh, *Mater. Res. Soc. Symp.* 670, K.2.2.1 (2001).
192. O. Sneh, R. B. Clark-Phelps, A. R. Londergan, J. Winkler, T. E. Seidel, *Thin Solid Films*, 402, 248 (2002).
193. J. W. Elam, Z. A. Sechrist and S. M. George, *Thin Solid Films* 414, 43 (2002).
194. S. Jeon, H. Yang, D.G. Park and H. Hwang, *Jpn. J. Appl. Phys.* 41, 2390 (2002).
195. D. Hausmann, J. Becker, S. Wang and R. G. Gordon, *Science* 298, 402 (2002).
196. K. Kukli, M. Ritala, M. Leskala, T. Sajavaara, J. Keinonen, D. Gilmer, S. Bagchi and L. Prabhu, *J. Non-Cryst. Sol.* 303, 35 (2002).
197. W. F. A. Besling, E. Young, T. Conard, C. Zhao, R. Carter, W. Vandervorst, M. Caymax, S. De Gendt, M. Heyns, J. Maes, M. Tuominen and S. Haukka, J. *Non-Cryst. Sol.* 303, 123 (2002).
198. D. M. Hausmann and R. G. Gordon, *J. Cryst. Growth* 249, 251 (2003).
199. H. Kim, P. C. McIntyre and K. C. Saraswat, J. *Mater. Res.* 19, 643 (2004).
200. D. R. G. Mitchell, D. J. Attard, K. S. Finnie, G. Triani, C. J. Barbe, C. Depagne and J. R. Bartlett, *Appl. Surf. Sci.* 243, 265 (2005).
201. Y. Kim and S. J. Yun, *J. Cryst. Growth* 274, 585 (2005).
202. A. Sherman, *US Patent Application* 20040221798 November 11, 2004.

5

Nitrides and Other Compounds

5.1 Introduction

Thin nitrides have been widely used in industry for many years. In the electronics industry, for example, silicon nitride has proven to be useful as a diffusion barrier to water vapor, as well as a good electrical insulator. Similarly, titanium nitride has been found to be an excellent diffusion barrier to oxygen, and is also a good electrical conductor. Also, the latter is a very hard material and has been widely used as a wear coating on tool steel.

In the microprocessor unit (MPU) interconnect technology Cu wiring has been generally utilized. The requirement for wiring is that the diffusion of copper into the adjacent dielectric must be prevented. In addition the interface has to be of high quality and suitable for copper. Tantalum nitride, titanium nitride, and tungsten nitride are among the most suitable barrier materials for this purpose. Molybdenum nitride can be used in tribological applications and its high conductivity has been utilized in microelectronics. It can also be applied in catalytic reactions.

5.2 Nitrides

5.2.1 Transition Metal Nitrides

5.2.1.1 Tantalum Nitride

Many studies have been made of TaN, due to the interest in using it as a barrier to diffusion of copper in integrated circuits [1–5].

The feasibility of growing tantalum nitride thin films was demonstrated quite early [1] using $TaCl_5$ and NH_3, although the phase obtained was Ta_3N_5 (a nitrogen rich dielectric). Since most applications in microelectronics require a diffusion barrier that is a good electrical conductor, and the TaN phase is such a conductor, the first issue that needed to be addressed was how to ensure that this would be the film formed. Secondly, any contamination of the film can be deleterious, depending on the application, so a process is necessary that would lead to a sufficiently pure TaN.

It was found that adding a pulse of zinc vapor as an additional reducing agent between each precursor pulse enabled cubic TaN to be formed [1] because NH_3 alone was not sufficient to reduce the Ta oxidation state to 3+. Some traces of zinc and chlorine remained in the deposited film, even at deposition temperatures as high as 450°C. In order to avoid the practical difficulties with the zinc process, two new nitrogen source reducing agents were tried [2]. They were tert-butylamine ($^tBuNH_2$) and allylamine (allylNH_2), with the hope that they would be more reactive than simply NH_3 and enable the growth of low resistivity TaN at a moderate temperature and with low levels of impurities. Although both reducing agents enabled ALD TaN, the allylamine produced very high resistivity films (> 18,000 μΩ-cm) with a considerable level of impurities at 400°C (8–10% chlorine, 10–14% carbon, 5–7% oxygen, and 4–7% hydrogen). This was even though NH_3 was added to the process.

Growth from the tert-butylamine reducing agent, when combined with NH_3, did yield somewhat better results. At 450°C, they obtained TaN films with resistivities between 2500 and 2700 μΩ-cm. Also, the carbon contamination was 6–8%, and the chlorine residual was 1.7–2.0%. As will be seen in the next chapter, the only practical way to grow low resistivity TaN with low levels of chlorine contamination at moderate temperatures is to use a plasma enhanced process.

If we limit our discussion to thermal processes, there are two other approaches that are being explored. One possibility would be to use an organometallic tantalum compound as the precursor. Two such compounds that have received some attention for a thermal ALD process with NH_3 as the reducing agent are: pentakisdimethylaminotantalum [PDMAT] [3] and tertbutylimidotris(diethylamido) tantalum [TBTDET] [4, 5].

When using PDMAT with NH_3 it was reported that films with some Ta and N could be grown, but that they contained a great deal of carbon, and were non-conducting [3]. In contrast, when TBTDET was used with NH_3 two other investigations reported that it was possible to produce electrically conducting films. In one case, it was reported that experiments carried out at 250°C grew 100% conformal amorphous films with a resistivity of 1000 µΩ-cm [4]. Also, the films had 10–20% oxygen, and the remainder was composed of Ta, C and N in the proportions 6:1:10. A later paper deposited a similar material from the same precursors [5] at about the same temperature. No measurements of resistivity or film composition were reported, but barrier quality was evaluated by sputtering copper onto the barrier film, and annealing at 500°C in N_2 for 30 minutes. For a barrier film as thin as 12 Å it was found that no evidence of etch pits caused by diffusion of copper was observed.

As has been outlined above, thermal ALD processes for TaN require rather a high temperature and there is a significant degree of contamination in the films from precursor material. Reactive-enhanced ALD processes have been used to attempt to overcome these issues. A good example is the case of TaN_x where the thermal reaction produces a sub-stoichiometric compound. By using a plasma-activated process, conductive cubic TaN with resistivity down to 380 µΩ can be deposited using $Ta(NMe_2)_5$ and H_2 as precursors at significantly lower temperatures [6, 7]. Figure 5.1 shows that long hydrogen plasma exposure times were necessary to achieve the lowest resistivity. It has also been demonstrated that the stoichiometry can be controlled by varying the gas mixture during the plasma step [8]. Figure 5.2 shows how the N/Ta atom ratio depends on the N_2/H_2 ratio. The material composition could be changed from $TaN_{0.3}$ to almost Ti_3N_5 and stoichiometric TaN could be obtained with $0.7 < P_{N2}/P_{H2} < 0.3$.

The effect of substrate temperature and time of plasma exposure on deposition rate per cycle and film stoichiometry was examined, and the results are shown in Figures 5.3 and 5.4.

There was a substantial temperature window between 300 and 400°C, and the stoichiometry seemed to be independent

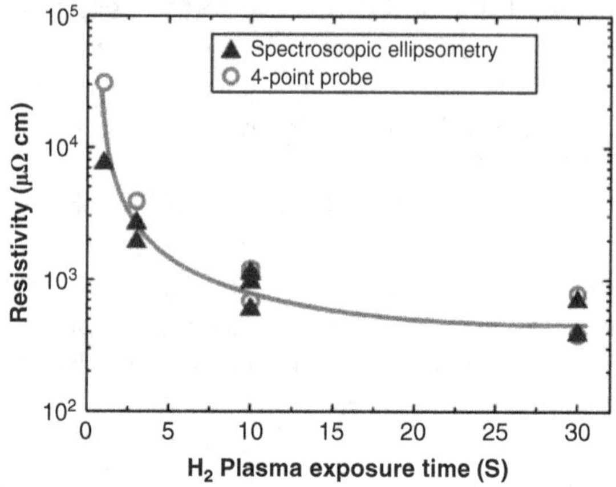

Figure 5.1 Room temperature resistivity of TaN films as a function of H_2 plasma exposure time [6]. *Reprinted from E. Langereis, H.C.M. Knoops, A.J.M. Mackus, F. Roozeboom, M.C.M. Van de Sanden and W.M.M. Kessels, J. Appl. Phys., 102, 083517, © 2007, American Institute of Physics.*

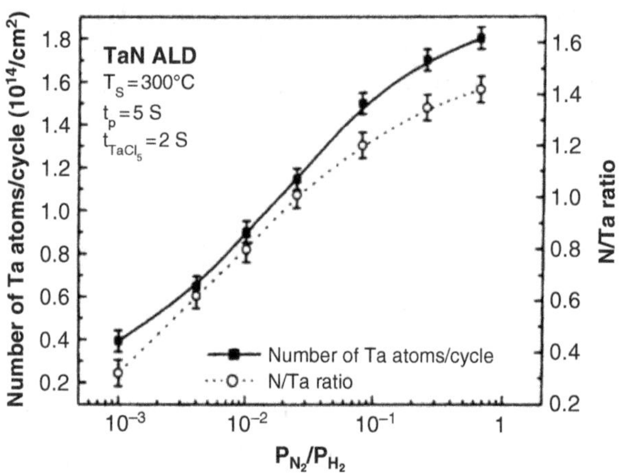

Figure 5.2 Deposition rate and stoichiometry of Ta_xN_y films grown at 300°C, plasma exposure of 5 s, and TaCl5 pulse of 2 s [8]. *Reprinted with permission from H. Kim, A. J. Kellock, S. M. Rossnagel, J. Appl. Phys. 92, 7080 © 2002, American Institute of Physics.*

of substrate temperature. As long as the plasma exposure time was 5 sec or longer, the deposition rate stabilized, and there was little influence on the 1:1 film stoichiometry (see Figure 5.3). The only impurities detected in these films, by RBS and FRES (forward recoil elastic spectrometry) were chlorine and hydrogen.

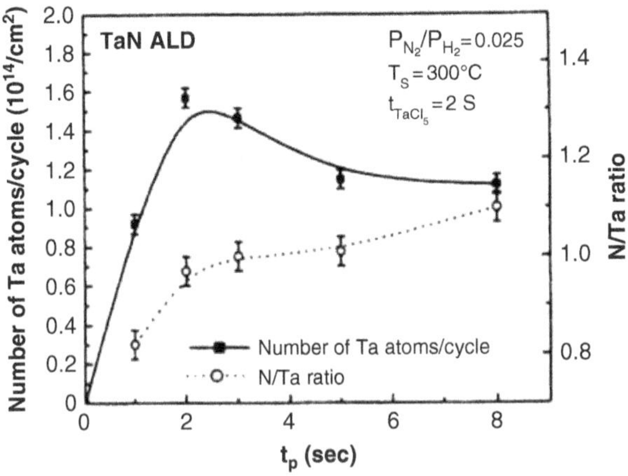

Figure 5.3 Deposition rate and stoichiometry vs. plasma exposure [8]. *Reprinted with permission from H. Kim, A. J. Kellock, S. M. Rossnagel, J. Appl. Phys. 92, 7080 © 2002, American Institute of Physics.*

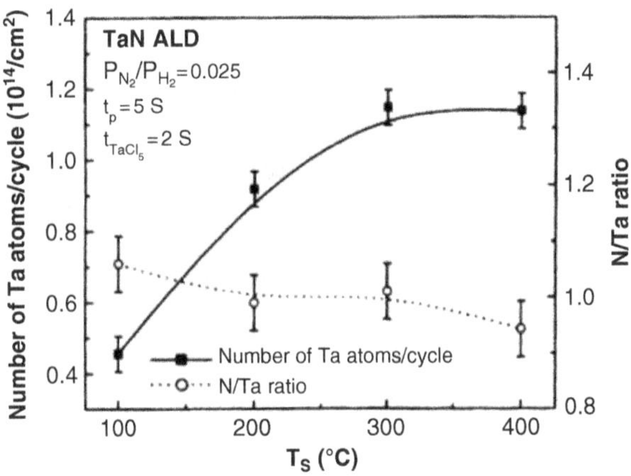

Figure 5.4 Deposition rate and stoichiometry vs. temperature [8]. *Reprinted with permission from H. Kim, A. J. Kellock, S. M. Rossnagel, J. Appl. Phys. 92, 7080 © 2002, American Institute of Physics.*

As long as the deposited films were capped with copper *in-situ*, no oxygen was detected. For plasma exposure times > 5 sec and substrate temperatures greater than 300°C, the chlorine content was less than 1%, however, the hydrogen content was on the order of 10%.

5.2.1.2 *Titanium Nitride*

Because of the commercial importance of TiN, it has been the most widely studied binary nitride film. Growth of TiN [9–22] by thermal ALD has been investigated using a variety of titanium bearing precursors and nitrogen bearing reducing agents. The most common titanium precursors studied, initially, were the halides ($TiCl_4$ and TiI_4), although the use of organometallic titanium compounds (i.e. tetrakis(dimethylamido)titanium, among others) has been examined more recently. The most frequently used reducing agent has been NH_3.

The first study of the deposition of TiN by ALD was intended to produce films that could be used to coat inexpensive soda lime glass [9] For this reason deposition was carried out at 500°C, using $TiCl_4$ as the titanium source, and NH_3 as the reducing agent. The overall reaction for these reactants was presumed to be

$$6TiCl_4(g) + 8NH_3(g) \rightarrow 6TiN(s) + 24HCl(g) + N_2(g) \quad (5.1)$$

Depositions were reported that used alternating pulses of $TiCl_4$ and NH_3, in a cross flow reactor using a nitrogen purge to separate the reactants. The chlorine impurity in the films was reported to be several atomic percent. Deposition rate was found to be quite low being 0.2–0.3 Å/cycle. Since deposition was only reported at one temperature, 500°C, the existence of a useful temperature window was not demonstrated. The measured film resistivity was found to be 250 μΩ-cm which is 10 times bigger than in the bulk material. In a later study [10] resistivity of 117 μΩ-cm was measured and the chlorine residue was 1 at-%.

Next, TiN films were grown in another reactor configuration ("vertical high vacuum chamber", and a "rotating injector" gas distributor) at pressures between 0.5 and 2 torr using argon as a purge gas [11]. In this case, depositions from 375–500°C were carried out, and as can be seen from Figure 5.5, there is a wide temperature window.

In addition, the variation of film resistivity with temperature is shown in Figure 5.6, where we see a resistivity of 200 μΩ-cm at 500°C. Again, this is a somewhat lower value than originally reported [9], but this would be consistent with the lower level of chlorine contamination (< 1%) being seen.

With the use of a plasma to increase the power of the reducing agent, however, we find that there is a great deal more process

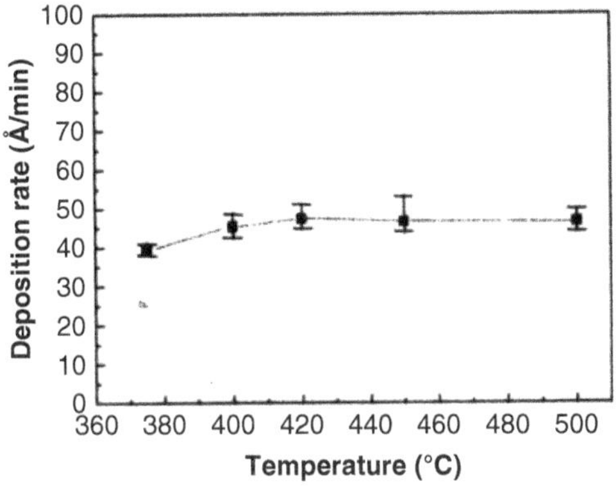

Figure 5.5 TiN deposition rate vs substrate temperature [11]. © *2003 The Japan Society of Applied Physics*

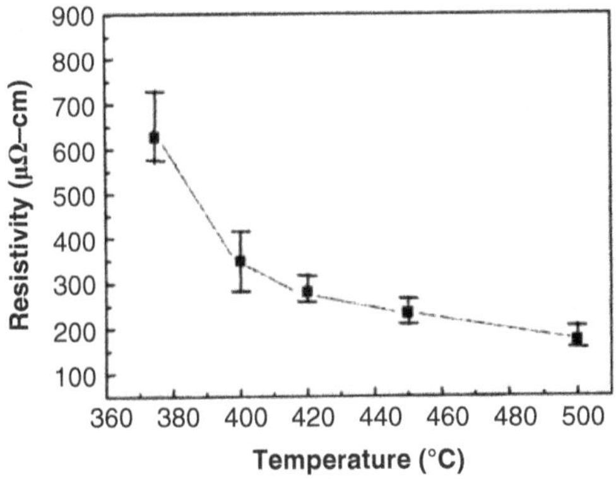

Figure 5.6 Film resistivity vs substrate temperature for ALD TiN [11]. © *2003 The Japan Society of Applied Physics.*

flexibility. Initial studies were done simply with $TiCl_4$ and NH_3 where a glow discharge was used to decompose the NH_3. The free radicals created by the discharge consisted of H, N, NH, and NH_2. Experiments were carried out in a downstream plasma REALD reactor, with the plasma source operated at ~ 250 watts. TiN films deposited using $TiCl_4$ as the metal source and a N_2/H_2 plasma have shown low impurity levels and low resistance compared with thermal ALD [23, 24].

Figure 5.7 shows that good resistivity can be obtained with much lower process temperature than for thermal ALD.

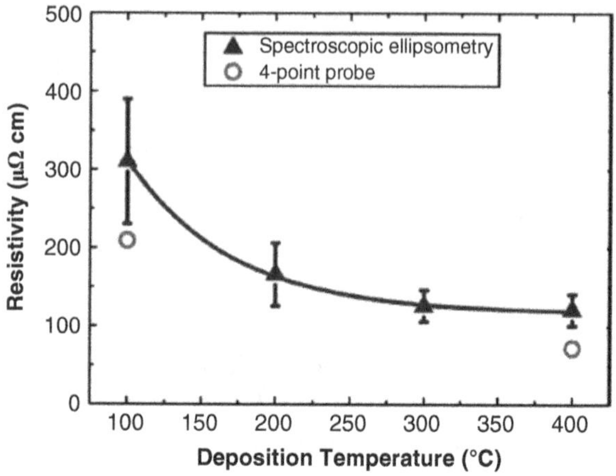

Figure 5.7 Room temperature resistivity of TiN films obtained by REALD [24] *Reprinted from J. Electrochem. Soc. 153, S. B. S. Heil, E. Langereis, F. Roozeboom, M. C. M. Van de Sanden, W. M. M. Kessels, Low-Temperature Deposition of TiN by Plasma-Assisted Atomic Layer Deposition, G956 © 2006, Reproduced by permission of ECS – The Electrochemical Society.*

Gold colored TiN films were grown at 150°C at a rate of 0.6 Å/cycle. They were adherent to the underlying silicon substrate, and the resistivity was ~ 200 μΩ-cm. Auger analysis of the film composition showed stoichiometric TiN with < 0.3% chlorine. A thin film of TiN deposited by this process into a high aspect ratio hole was very conformal, and is shown in Figure 5.8. In order to clarify the mechanisms underlying these experiments, beam studies of the chemisorption of $TiCl_4$, and the abstraction of chlorine from the adsorbed species, were performed in a beam apparatus at UC Berkeley [21]. This study showed that $TiCl_4$ chemisorbed dissociatively, even at room temperature, as $TiCl_2$. It was also confirmed that hydrogen atoms were very effective at abstracting chlorine from the adsorbed $TiCl_2$ monolayer.

In addition to the halides of titanium, there are other titanium organometallic precursors containing nitrogen that can be used to grow TiN by REALD. Their advantage is that they do not introduce any chlorine contamination, but they do produce films with some level of carbon contamination. Chlorine contamination may be undesirable when these TiN films are used in integrated circuits where aluminum is used as a conductor, because of the potential formation of aluminum chloride. On the other hand, the presence of small amounts of carbon does not raise any

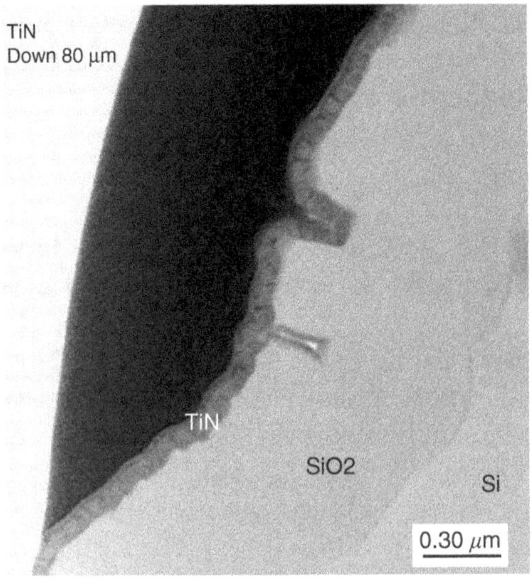

Figure 5.8 TEM of REALD TiN film showing conformality.

corrosion issues. A convenient titanium organometallic precursor is tetrakis(dimethylamino)titanium (TDMAT), and studies have been reported where it has been reduced by either using a nitrogen plasma, a hydrogen plasma, a plasma created in a hydrogen/nitrogen mixture, or an ammonia plasma [25–27].

Experiments were carried out in a downstream REALD system operated at 13.56 MHz and between 50 to 300 watts. Argon was used as a purge gas, and the temperature range was 150 to 350°C. The TDMAT vapor was evaporated in a 45°C external reservoir, and introduced with an argon carrier gas. All the films grown exhibited a temperature window extending from 200 to 300°C. The observed growth rate for this process ranged from 2 to 4 Å/cycle. Film resistivity values ranged from 300 µΩ-cm to 1500 µΩ-cm, which is on the high side for good quality TiN. In general, these films were found to be nitrogen rich, and contained about 3 to 15% carbon and oxygen. Growth of a film over a 10:1 aspect ratio hole in silicon showed almost 100% conformality.

Examination of one 170 Å thick film by XTEM showed it to be partially crystallized with a microcrystalline structure. A diffusion barrier test was performed by depositing test samples composed of Cu/TiN/SiO_2/Si, and heat treating them from 400 to 600°C for 1 hour. Films grown at 250°C from nitrogen plasma were found to perform well as diffusion barriers, up to 450°C. At temperatures of

500°C and above, there was clear evidence of Cu_3Si formation, indicating mixing of the Si and Cu. Presumably, these films are crystallizing at these higher temperatures.

5.2.1.3 Other Transition Metal Nitrides

Other transition metal nitrides which have been deposited are ZrN, NbN, MoN, HfN and WN, both by thermal and plasma-enhanced methods.

It has been reported [28] that a process using $NbCl_5$ and NH_3 at 500°C produced a polycrystalline NbN (1:1) film with a resistivity of 600 μΩ-ohm. Films several 1000Å thick were smooth, and were deposited at a rate of 0.25 Å/cycle. Adding a zinc vapor pulse produced similar films, reduced the resistivity by a factor of two, but also reduced the deposition rate. Interestingly, no chlorine residue (<0.3%) was seen in an RBS scan for films with or without zinc. Lower temperature deposition between 250 and 300°C was achieved using $Nb(N^tBu)(NEtMe)_3$ and H_2 or H_2/N_2 plasma [29].

The first investigation of ALD of tungsten nitride [30], used WF_6 and NH_3 as precursors. Depositions were carried out in the temperature range between 325–525°C, and self limiting surface reactions were confirmed. The film composition was reported to be approximately W_2N, and although they were smooth, their resistivity was very high 4.5 Ω-cm. As well, the film conformality was not reported, and no information on possible fluorine contamination was given. A subsequent study looked at deposition from these same precursors [31] at 350°C, and found films with a W:N ratio of 1.62. In this case, the fluorine content was reported to be 2.4 at%. However, the film resistivity was 4500 μΩ-cm, which is many orders of magnitude less than in the previous study.

Aside from the unacceptably high resistivity, the WF_6 process has the further disadvantage of fluorine contamination which can attack either silicon or silicon dioxide. The obvious way to eliminate any concern about fluorine contamination would be to use an organometallic precursor. A precursor with the composition Bis(tert-butylimido)bis(dimethylamido)tungsten $(^tBuN)_2(Me_2N)_2W$ was prepared [32] and it was reported to have a vapor pressure of 0.037 torr at 30°C, and presumably was thermally stable at this temperature. Using this compound with NH_3 at 300–350°C, produced thin films of conformal amorphous $WN_{1.1}$, with less than 0.5% carbon contamination and a resistivity of 1500–4000 μΩ-cm. Also noted was the fact that resistivity could be reduced further by annealing. As expected, the films were all very conformal on very high aspect ratio holes.

WN has also been deposited by ALD from WF_6 and NH_3 plasma at 350°C and behaved as an effective barrier to copper diffusion at 600°C [33].

Molybdenum nitride has been grown by thermal ALD using either chloride or alkylamide precursors [9, 15, 34, 35]. The temperature range necessary using clorides was 350–500°C while alkylamides were successful in growing films at lower temperatures of 260–300°C. These gave a variety of crystal structures : amorphous, cubic, hexagonal and tetragonal.

As was noted in our earlier discussion of ALD of tantalum nitride, there can be more than one stable phase of a solid compound. In a similar fashion, since there is some interest in the deposition of dielectric hafnium and zirconium nitrides, a process has been demonstrated [36] that enables Hf_3N_4 and Zr_3N_4 to be deposited from tetrakis(dimethylamido)hafnium $Hf(NMe_2)_4$ and tetrakis(dimethylamido)zirconium $Zr(NMe_2)_4$ along with NH_3. In this case, the nitrides were found to be dielectrics with high dielectric constants – 30 for Hf_3N_4 and 36 for Zr_3N_4. ALD was carried out within a temperature window, between 150–250°C, and the films were stoichiometric, smooth, and amorphous. Plasma ALD of ZrN and HfN has also been achieved, again using plasma activated H_2, H_2/N_2 or NH_3 [37–41].

5.2.2 Group III Nitrides

5.2.2.1 *Aluminum nitride*

Aluminum nitride is an interesting material, and has found use in a number of applications. It is an excellent thermal conductor, is a good electrical insulator and is generally stable in oxidizing environments. It is difficult to deposit by thermal ALD, but it can be readily deposited at moderate temperatures by REALD [42].

Recently, deposition was carried out in a parallel plate capacitively coupled reactor with a 13.56 MHz RF power supply using trimethylaluminum and ammonia as precursors with a N_2 carrier gas [44]. The temperature range for well-controlled growth was 150–300°C as shown in Figure 5.9.The H concentration was quite high, varying from 27 at.% at 150°C to 13 at.% at 300°C however the C and O impurity levels were low 2.3–2.5 at.% and 0.07–0.25 at.% respectively. The AlN was shown to be an effective passivation cap for GaAs devices.

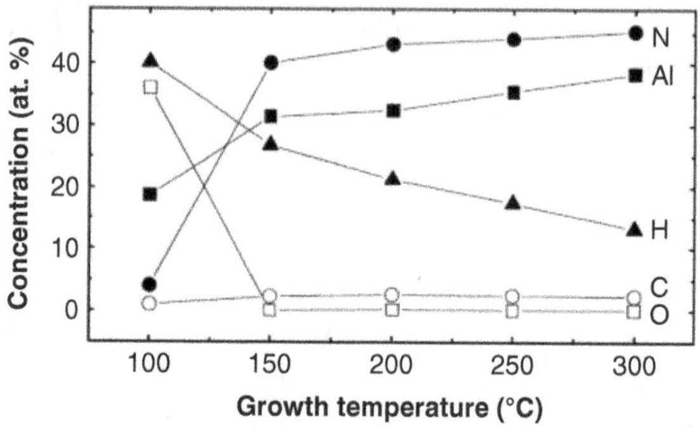

Figure 5.9 Atomic concentration in AlN as a function of the ALD growth temperature [44]. *Reprinted from Appl. Surf. Sci. 257, M. Bosund, T. Sajavaara, M. Laitinen, T. Huhtio, M. Putkonen, V. M. Airaksinen, and H. Lipsanen, Properties of AlN grown by plasma enhanced atomic layer deposition, 7827© 2011 with permission from Elsevier.*

A number of other studies have looked at the possibility of depositing AlN by thermal ALD. The precursors examined were:

dimethylethylamine alane (DMEAA) + ammonia (NH_3) [45, 46]

dimethylethylamine alane (DMEAA) + dimethylhydrazine (DMHy) [42]

trimethylaluminum (TMA) + ammonia (NH_3) [47]

Although these studies only examined a few ALD cycles, it was apparent that these particular processes were not going to lead to commercially useful processes. The difficulty with DMEAA was that it was sensitive to thermal decomposition at low temperatures (150°C), and at this temperature neither of the nitriding agents (NH_3, DMHy) were sufficiently reactive to complete the surface reaction in a reasonable time. In other words, there was no temperature window available. When TMA was used, by contrast, it was stable enough, but the reaction with NH_3 left films with excess carbon, so that pure AlN could not be formed.

5.2.2.2 Boron Nitride

Boron nitride can also be deposited by thermal ALD from boron trichloride [48] or boron tribromide [49], using ammonia as the second reactant. In the study using BBr_3, deposition was carried out at just two temperatures 400°C and 750°C, and the deposition rate appeared to be constant at 0.45 Å/cycle. At the lower temperature, the film grown was amorphous, while at the higher temperature it had a turbostratic structure. They were also close to stoichiometric (B/N = 0.95), and carbon levels were on the order of 2–3%. If ALD at 750°C is mandated, then this will be a difficult reactor to build and maintain. Using BCl_3 and NH_3 at 225°C [48], the as grown films were stoichiometric and amorphous.

5.2.2.3 *Other Group III Nitrides*

The semiconductors such as GaN, InN, etc. are referred to in Chapter 3 and will not be covered further here.

5.2.3 Group IV Nitrides

Of the group IV nitrides, silicon nitride is the one which has had significant work done on it.

In one example, Si_2Cl_6 was used as the silicon source, hydrazine (N_2H_4) was the nitrogen bearing reducing agent [50], and stoichiometric Si_3N_4 was deposited. No chlorine contamination was seen on Auger analysis. Deposition was constant over a temperature window from 525–650°C, and the rate was 2.3 Å/cycle which corresponds to a monolayer of film per deposition. However, the length of each hydrazine exposure was exceptionally long. The explanation for this long time exposure lies in the fact that a constant 1 ML/cycle deposition rate did not occur until the N_2H_4 exposure exceeded 10^8 L. For a reactor pressure of 0.10 torr, this requires a ~ 1000 sec exposure per cycle. The ALD cycle reported is illustrated in Table 5.1.

Adding up all the times, we observe that each cycle takes ~ 13 minutes. Therefore, we have a deposition rate of 2.3/13 or 0.177 Å/min. At this rate a 200Å thick film would require a 35 minute deposition. It would seem unlikely that such a slow deposition rate would enable a commercially viable process.

A similar process [51] was explored using $SiCl_4$ and NH_3 with similar results. Over a process temperature of 425–625°C and a reactor pressure of 10 torr the deposition rate was constant at 2.45 Å/cycle, as long as the reactant exposure exceeded 10^{10} L. This again would lead to a 1000 sec exposure of reactant per cycle.

Plasma ALD of SiN_x has also been reported [52].

Table 5.1 Si_3N_4 ALD parameters for a typical cycle [50]. *Reprinted from Appl. Surf. Sci. 112, S. Morishita, S. Sugahara and M. Matsumura, Atomic-layer chemical-vapor-deposition of silicon-nitride,198, © (1997) with permission from Elsevier.*

Chamber temperature	600°C
Si_2Cl_6 exposure time, pressure	15sec, 1 torr
N_2 purge	7 sec, 3 times
Evacuation time	10 sec
N_2H_4 exposure time, pressure	720 sec, 0.1 torr
Evacuation time	10 sec
N_2 purge	7 sec, 3 times

5.2.4 Mixed Nitrides

In the preceding sections of this chapter, we have seen that we can deposit, for example, TiN or AlN by thermal or plasma ALD. It would then be reasonable to expect that we could deposit films containing varying percentages of these materials by simply alternating several different deposition sequences. With the goal of creating a diffusion barrier that will improve on the well known TiN, two studies examined how such a ternary nitride could be deposited. In one case [53], a layer of TiN was deposited in a first subcycle, from $TiCl_4$ and a H_2/N_2 plasma, and this was followed by one or more subcycles for the deposition of a layer of AlN from $AlCl_3$ and a H_2/NH_3 plasma. The deposition was carried out in a capacitively coupled RF powered (150 watts at 13.56 MHz) reactor at 350°C. For the TiN cycles, the sequence for metal precursor-purge-plasma-purge was 1-15-10-5 s and for the AlN cycles it was 3-10-10-5 s. For the times cited, the growth rates of each film cycle were in the fully saturated regime, and were ~ 0.5 Å/cycle. The titanium content of the films was changed by adding multiple TiN subcycles for each AlN cycle. In the films grown, chlorine and hydrogen concentrations were found to be 0.6 and 1.5 %. In addition, the film resistivity varied from 80 μΩ-cm for TiN to 300 μΩ-cm for films with 20 % aluminum. Films with 17 % aluminum were largely polycrystalline with amorphous regions. The hope was that these amorphous regions would provide superior diffusion barrier characteristics.

A second study from the same group [54], used organometallic precursors. For the TiN film TDMAT and NH_3 are used with a hydrogen plasma. The aluminum step used an exposure to TMA and then again a hydrogen plasma. This sequence is illustrated graphically in Figure 5.10.

For this scheme, deposition was carried out at 180°C, and a plasma power of 200 watts. It was demonstrated that film thickness was a linear function of the number of cycles, with no evidence of delayed nucleation. The growth of one TiN cycle was 2.4 Å, and for one Al cycle it was 1.5 Å. The overall growth rate of the Ti-Al-N film was 3.5 Å/cycle.

Films were grown on SiO_2 coated silicon wafers, and were polycrystalline and fairly smooth (3.32 Å rms vs. 2.50 Å rms for original SiO_2 surface). XRD analysis indicated that the film was made up of a mixture of $Ti_3Al_2N_2$ and TiN. As expected, the films were conformal, and the resistivity was ~ 500 μΩ. This high a value could be due to the carbon (~ 10 atom %), and the hydrogen (~ 5 atom %) content.

Resistance of this film due to oxidation, in an oxygen annealing environment, was compared to that of a TiN film. Annealing

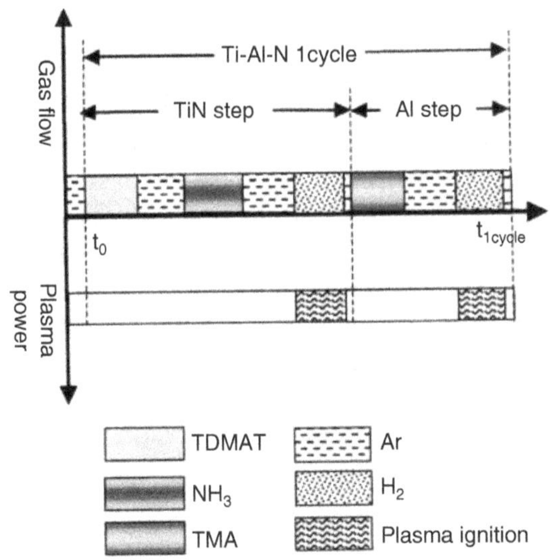

Figure 5.10 Two part cycle to deposit Ti-Al-N films by REALD [54]. *Reprinted from Electrochem. Solid-State Lett. 6, Y. J. Lee, S. W. Kang, Ti-Al-N Thin Films Prepared by the Combination of Metallorganic Plasma-Enhanced Atomic Layer Deposition of Al and TiN, C70 © 2003, Reproduced by permission of ECS – The Electrochemical Society.*

was carried out on 500 Å thick films at 700°C. The TiN film exhibited a greater than 30 % oxygen content distributed uniformly after annealing, while the Ti-Al-N film showed oxygen concentrated primarily near the film surface.

The deposition of Ti_xSi_yN have been reported by plasma ALD using $TiCl_4$, SiH_4 and $N_2/H_2/Ar$ radicals at a substrate temperature of 350°C [55]. The sequence was $TiCl_4$ pulse, SiH_4 pulse then $N_2/H_2/Ar$ plasma and a deposition rate of 0.73 Å/cycle with a Si content of 8.5 at.% was achieved. There was 0.5 at.% Cl contamination in the films. The addition of Si to the films resulted in increased resistivity (~500 μΩ cm) and the diffusion barrier properties were better than for Si_3N_4.

5.3 Chalcogenides

Metal chalcogenides have a long history in ALD. The early work by Suntola on zinc sulphide resulted in an early patent [56] In this case, the precursor were elemental Zn and S evaporated in directed beams which struck a rotating substrate. The impetus for this work was the development of thin film electroluminescent display materials. For more efficient production, materials such as $ZnCl_2$ and H_2S proved to be easier to work with [57] and all subsequent work has been based on compound precursors. ZnS has a bandgap above the visible region so in order to produce visible luminescence, the ZnS has to be doped with another element which will provide lower energy photon emission. Typically rare earth metals and alkaline earth metals are used as dopants so there have been a number of studies on these materials, for example [58–63]. Deposition of other sulphides has also been carried out for their use as buffer layers in photovoltaic cells, particularly CdS and InS [64, 65].

A general problem with halide precursors is the existence of a small amount of Cl in the subsequent film which can have an effect on its performance. To avoid this, non-halide precursors were developed, similar to those used for other metal compounds. Typical examples of these are zinc acetate [66] and $Zn(thd)_2$ (thd = 2,2,6,6-tetramethyl-3,5-heptanedionate) [67] It is more convenient to have precursors which are liquid at room temperature and do not require much (if any) heating to achieve a high enough vapour pressure so diethyl zinc (DEZ) has become widely used [68] and dimethyl zinc (DMZ) has also been studied [69].

The process temperature for ZnS deposition is very dependent on the precursor. Halide precursors typically have deposition temperatures in the range 350–500°C. For example, at 500°C substrate temperature using $ZnCl_2$ and H_2S gave a deposition rate of ~1 Å/cycle with the stabilization time when this growth rate was achieved varying depending on substrate as shown in Figure 5.11 [70]

A lower deposition rate of ~0.4 Å/cycle was reported using ZnI_2 as the metal precursor but this rate was approximately constant over the temperature range 300 to 500°C as shown in Figure 5.12 [71]. The

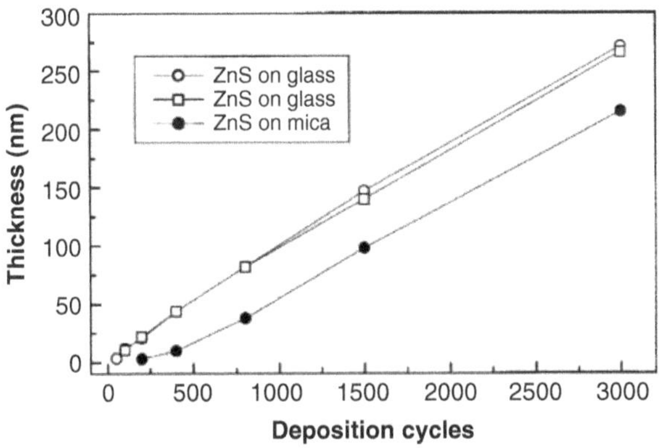

Figure 5.11 Film thickness dependence on number of deposition cycles for various substrates [70]. *Reprinted from Appl. Surf. Sci. 120, J. Ihanus, M. Ritala, M. Leskelä, T. Prohaska, R. Resch, G. Friedbacher, M. Grasserbauer, AFM studies on ZnS thin films grown by atomic layer epitaxy, 43 © 1997, with permission from Elsevier.*

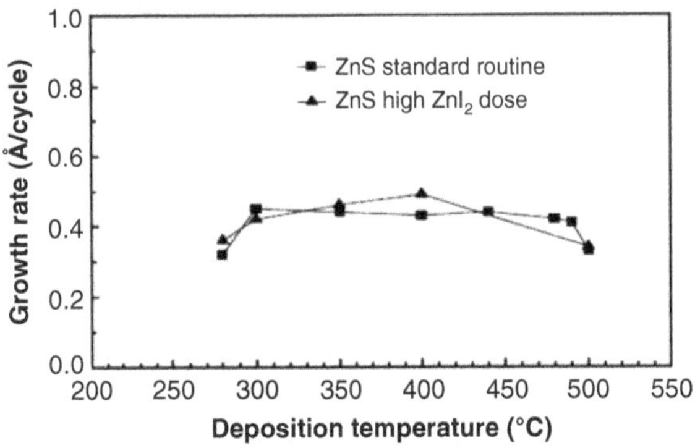

Figure 5.12 Growth rate of ZnS thin films as a function of the deposition temperature. Adapted from [71].

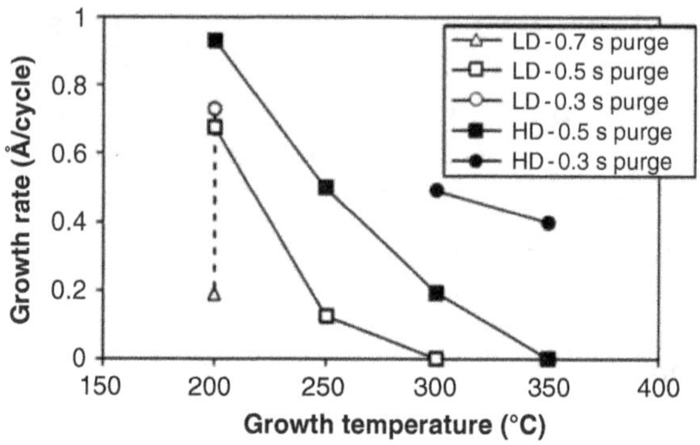

Figure 5.13 ZnS growth rate as a function of growth temperature, under different DEZn dosing and purge times conditions [72]. *Reprinted from J. Cryst .Growth 234, G. Stuyven, P. De Visschere, A. Hikavyy, K. Neyts, Atomic layer deposition of ZnS thin films based on diethyl zinc and hydrogen sulfide, 690 © 2002 with permission from Elsevier.*

difference in growth rate is due to the larger size of the iodide molecule and which reduced zinc coverage of the substrate.

Using DEZ as a precursor can enable deposition to take place at much lower temperatures, even down to room temperature. However, there is a large dependence of growth rate with temperature as shown in Figure 5.13 [72] and growth ceases at higher temperatures.

ZnS can form in two phases, the zinc blende, cubic, phase and the wurtzite, hexagonal, phase. Where polycrystalline films have been deposited and the crystal structure of the films has generally found to be the cubic form at lowered temperature with some evidence for the hexagonal phase at higher deposition temperatures. A very comprehensive review of the crystallographic measurements on a wide range of ALD materials, including chalcogenides, has recently been published. The reader is referred to this for more detailed information [73].

Some work has been carried out on mixed chalcogenide films. Strained layer superlattices with nanolaminate ZnS and ZnSe structures were deposited and it was seen that the luminescence peak position varied as the width of the ZnSe quantum well was varied [74]. Other work on ternary chalcogenides such as ZnInS has been carried out for their possible use as buffer layers in CIGS solar cells.

Interest has recently been rekindled based in a new class of chalcogen precursors – alkylsilyl compounds [75, 76], for example

$(Et_3Si)_2Te$ for the deposition of phase-change memory materials such as $Ge_2Sb_2Te_5$.

5.4 Other Compounds

There have been a few other materials which have been studied to a smaller extent, mainly halides.

Metal fluorides are important materials for optical applications such as filters because of their low refractive indices. however, there has been limited work on ALD of fluorides. The obvious F precursor, HF, poses difficulties due to its extreme corrosivity and toxicity. In some early studies, HF was generated as an intermediate step in the reaction of NH_4F vapour with zinc acetate, $Ca(thd)_2$ or $Sr(thd)_2$ [77]. Other fluorine compounds such as TiF_4 and TaF_5 have been used as the fluorine precursor when combined with $Ca(thd)_2$ [78, 79], $Mg(thd)_2$ [80], $La(thd)_3$ [81] or $Y(thd)_3$ [82]. However, traces of the metal from the fluoride precursor were found in the deposited film and this contamination increased as the deposition temperature was reduced in an attempt to decrease the film roughness.

One other example is the wide bandgap semiconductor CuCl which is of interest for optical devices because of its high exciton binding energy and close lattice match to silicon. ALD has been used to deposit arrays of quantum dots of this material using [bis(trimethylsilyl)acetylene](hexafluoro-acetylacetonato)-copper(I) and HCl [83].

References

1. M. Ritala, P. Kalsi, D. Riihelä, K. Kukli, M. Leskelä and J. Jokinen, *Chem. Mater.*, 11, 1712 (1999).
2. P. Alen, M. Juppo, M. Ritala, M. Leskelä, T. Sajavaara and J. Keinonen, *J. Mater. Res.* 17, 107 (2002).
3. H. Wojcik, M. Friedemann, M. Feustel, M. Albert, S. Ohsiek, J. Metzger, J. Voss, J.W. Bartha and C. Wenzel, *International Interconnect Technology Conference, IEEE*, 19 (2007).
4. D. Cheng and E. T. Eisenbraum, *Mat. Res. Soc. Symp. Proc.* 766, E10.4.1 (2003).
5. K. Kim, T. J. Anderson, and L. McElwee-White, *Annual AIChE Mtg.* Cincinnati, Ohio #566 (2005).
6. E. Langereis, H.C.M. Knoops, A.J.M. Mackus, F. Roozeboom, M.C.M. Van de Sanden and W.M.M. Kessels, *J. Appl. Phys* 102, 083517 (2007).

7. H. Kim, C. Detavernier, O. Van der Straten, S. M- Rossnagel, A.J. Kellock and D.G. Park, J. Appl. Phys. 98, 014308 (2005).
8. H. Kim, A. J. Kellock and S. M. Rossnagel, *J. Appl. Phys.* 92, 7080 (2002).
9. L. Hiltunen, M. Leskelä, M. Mäkelä, L. Niinistö, E. Nykänen and P. Soininen, *Thin Solid Films* 166, 149 (1988).
10. J. Uhm, S. Lee, J. Lee, Y-D. Kim and H. Jeon, *J. Korean Phys. Soc.* 35, S765 (1999).
11. J. Kim, H. Hong, S. Ghosh, K-Y. Oh and C. Lee, *Jpn. J. Appl. Phys.* 42, 1375 (2003).
12. D-G. Park, K-Y. Lim, H-J. Cho, T-H. Cha, I-S. Yeo, J-S. Roh and J. W. Park, *Appl. Phys. Lett.* 80, 2514 (2002).
13. M. Ritala, M. Leskelä, E. Rauhala and P. Haussalo, *J. Electrochem. Soc.* 142, 2731 (1995).
14. M. Ritala, T. Asikainen, M. Leskelä, J. Jokinen, R. Lappalainen, M. Utrainen, L. Niinistö and E. Ristolainen, *Appl. Surf. Sci.* 120, 199 (1997).
15. M. Juppo, M. Ritala and M. Leskelä, *J. Electrochem. Soc.* 147, 3377 (2000).
16. M. Juppo, P. Alen, M. Ritala, T. Sajavaara, J. Keinonen and M. Leskelä, *Electrochem and Solid-State Ltrs.* 5, C4 (2002).
17. M. Ritala, M. Leskelä, E. Rauhala and J. Jokinen, *J. Electrochem. Soc.* 145, 2914 (1998).
18. P. Mårtensson, M. Juppo, M. Ritala, M. Leskelä, and J-O. Carlsson, J. *Vac. Sci. Technol.* B17, 2122 (1999).
19. J-S. Min, Y-W. Son, W-G. Kang, S-S. Chun and S-W. Kang, *Jpn. J. Appl. Phys.* 37, 4999 (1998).
20. J. Y. Kim, G. H. Choi, Y. D. Kim, Y. Kim and H. Jeon, *Jpn. J. Appl. Phys.* 42, 4245 (2003).
21. S. G. Park and D-H. Kim, *Jpn. J. Appl. Phys.* 43, 303 (2004).
22. J-H. Yun, E-S. Choi, C-M. Jang and C-S. Lee, *Jpn. J. Appl. Phys.* 41, L418 (2002).
23. K. E. Elers, J. Winkler, K. Weeks, and S. Marcus, *J. Electrochem. Soc.* 152, G589 (2005).
24. S. B. S. Heil, E. Langereis, F. Roozeboom, M. C. M. Van de Sanden, and W. M. M. Kessels, *J. Electrochem. Soc.* 153, G956 (2006).
25. J. Y. Kim, Y. Kim and H. Jeon, *Jpn. J. Appl. Phys.* 42, L415 (2003).
26. J. Y. Kim, S. Seo, D. Y. Kim, H. Jeon and Y. Kim, *J. Vac. Sci. Technol.* 22, 8 (2004).
27. J. Y. Kim, D. Y. Kim, H. O. Park and H. Jeon, *J. Electrochem. Soc.* 152, G29 (2005).
28. K-E. Elers, M. Ritala, M. Leskelä and E. Rauhala, *Appl. Surf. Sci.* 82/83, 468 (1994).
29. J. Hinz, A. J. Bauer, T. Thiede, R. A. Fischer, and L. Frey, *Semicond. Sci. Technol.* 25, 075009 (2010).
30. J. W. Klaus, S. J. Ferro and S. M. George, *J. Electrochem. Soc.* 147, 1175 (2000).

31. K-E. Elers, V. Saanila, P. J. Soininen, W-M. Li, J. T. Kostamo, S. Haukka, J. Juhanoja and W. F. A. Besling, *Chem. Vap. Deposition* 8, 149 (2002).
32. J. S. Becker and R. G. Gordon, *Appl. Phys. Lett.* 82, 2239 (2003).
33. H. S. Sim, S. I. Kim, H. Jeon, and Y. T. Kim, *Jpn. J. Appl. Phys.*, Part 1 42, 6359 (2003).
34. V. Miikkulainen, M. Suvanto, and T. Pakkanen, *Chem. Mater.* 19, 263 (2007).
35. V. Miikkulainen, M. Suvanto, and T. Pakkanen, *Chem. Vap. Deposition* 14, 71 (2008).
36. J. S. Becker, E. Kim and R. G. Gordon, *Chem. Mater.* 16, 3497 (2004).
37. S. Consiglio, W. X. Zeng, N. Berliner, and E. T. Eisenbraun, *J. Electrochem. Soc.* 155, H196 (2008).
38. W. J. Maeng, G. H. Gu, C. G. Park, K. Lee, T. Lee, and H. Kim, *J. Electrochem. Soc.* 156, G109 (2009).
39. E. J. Kim and D. H. Kim, Electrochem. *Solid-State Lett.* 9, C123 (2006).
40. S. Cho, K. Lee, P. Song, H. Jeon, and Y. Kim, *Jpn. J. Appl. Phys.*, Part 1 46, 4085 (2007).
41. S. E. Potts, C. J. Carmalt, C. S. Blackman, F. Abou-Chahine, N. Leick, W. M. M. Kessels, H. O. Davies, and P. N. Heys, *Inorg. Chim. Acta* 363, 1077 (2010).
42. D. W. Robinson and J. W. Rogers, Jr., *Thin Solid Films* 372, 10 (2000).
43. Y. J. Lee and S. W. Kang, *Thin Solid Films* 446, 227 (2004).
44. M. Bosund, T. Sajavaara, M. Laitinen, T. Huhtio, M. Putkonen, V. M. Airaksinen, and H. Lipsanen, *Appl. Surf. Sci.* 257, 7827 (2011).
45. J. N. Kidder, Jr., H. K. Yun, J. W. Rogers, Jr. and T. P. Pearsall, *Chem. Mater.* 10, 777 (1998).
46. H. Liu and J. W. Rogers, Jr., *J. Vac. Sci. Technol.* A17, 325 (1999).
47. R. L. Puurunen, A. Root, P. Sarv, S. Haukka, E. I. Iiskola, M. Lindblad and A. O. I. Krause, *Appl. Surf. Sci.* 165, 193 (2000).
48. J. D. Ferguson, A. W. Weimer and S. M. George, *Thin Solid Films* 413, 16 (2002).
49. B. Marlid, M. Ottosson, U. Pettersson, K. Larsson and J-O. Carlsson, *Thin Solid Films* 402, 167 (2002).
50. S. Morishita, S. Sugahara and M. Matsumura, *Appl. Surf. Sci.* 112, 198 (1997).
51. J. W. Klaus, A. W. Ott, A. C. Dillon, and S. M. George, *Surf. Sci.* 418, L14 (1998).
52. T. K. Eom, S. H. Kim, K. S. Park, S. Kim, and H. Kim, Electrochem. *Solid-State Lett.* 14, D10 (2011).
53. Y. J. Lee and S. W. Kang, J. *Vac. Sci. Technol.* A21, L13 (2003).
54. Y. J. Lee and S. W. Kang, Electrochem. *Solid-State Lett.* 6, C70 (2003).
55. J. S. Park and S. W. Kang, Electrochem. *Solid-State Lett.* 7, C87 (2004).
56. T. Suntola and J. Antson, *Finnish Patent* 52 359, (1977).
57. T. Suntola, J. Antson, A. Pakkala, and S. Lindfors, in SID International Symposium in San Diego, California, 29 April-1 May 1980, Digest of Technical Papers (SID, Los Angeles, California, 1980) pp. 108–109.

58. S. Tanaka, J. Cryst. Growth 101, 958 (1990).
59. M. Tammenmaa, H. Antson, M. Asplund, L. Hiltunen, M. Leskelä, L. Niinistö, and E. Ristolainen, *J. Cryst. Growth* 84, 151 (1987).
60. P. Soininen, E. Nykänen, L. Niinistö, and M. Leskelä, *Chem. Vap. Deposition* 2, 69 (1996).
61. M. Tammenmaa, M. Leskelä, T. Koskinen, and L. Niinistö, J. Less-*Common Met.* 126, 209 (1986).
62. M. Leskelä and M. Tammenmaa, *Mater. Chem. Phys.* 16, 349 (1987).
63. S. Dey and S J Yun, *Applied Surface Science* 143, 191(1999).
64. J. R. Bakke, H. J. Jung, J. T. Tanskanen, R. Sinclair, and S. F. Bent, *Chem. Mater.*, 22, 4669 (2010).
65. E.B. Yousfi, B. Weinberger, F. Donsanti, P. Cowache, D. Lincot, *Thin Solid Films* 387, 29 (2001).
66. K. Kobayashi and S. Okudaira, *Chem. Lett.* 26, 511 (1997).
67. M. Leskelä and L. Niinistö, in *Atomic layer epitaxy*, edited by T. Suntola and M. Simpson (Blackie and Son, London, 1990) pp. 1–39.
68. Y.T. Lin, P. H. Chung, H. W. Lai, H. L. Su, D. Y. Lyu, K. Y. Yen, T. Y. Lin, C. Y. Kung, J. R. Gong, *Appl. Surf. Sci.* 256, 819 (2009).
69. A. Hunter and A. H. Kitai, *J. Cryst. Growth* 91, 111 (1988).
70. J. Ihanus, M. Ritala, M. Leskelä, T. Prohaska, R. Resch, G. Friedbacher, M. Grasserbauer, *Applied Surface Science* 120, 43 (1997).
71. J. Ihanus, PhD Thesis, Helsinki University, 2010.
72. G. Stuyven, P. De Visschere, A. Hikavyy, K. Neyts, J. *Cryst. Growth* 234 690 (2002).
73. V. Miikkulainen, M. Leskelä, M. Ritala and R. Puurunen, *Appl. Phys. Rev.* 113, 021301 (2013).
74. M. Yokoyamaa, N. –T. Chen, *J. Cryst. Growth* 223, 369 (2001).
75. K. Knapas, T. Hatanpää, M. Ritala, and M. Leskelä, *Chem. Mater.*, 22, 1386 (2010).
76. V. Pore, T. Hatanpää, Mikko Ritala, and Markku Leskelä, J. *Am. Chem. Soc.*, 131,3478 (2009).
77. M. Ylilammi and T. Ranta-aho, *J. Electrochem. Soc.* 141, 1278 (1994).
78. T. Pilvi, K. Arstila, M. Leskel¨a, and M. Ritala, *Chem. Mater.* 19, 3387 (2007).
79. T. Pilvi, E. Puukilainen, U. Kreissig, M. Leskelä, and M. Ritala, *Chem. Mater.* 20, 5023 (2008).
80. T. Pilvi, M. Ritala, M. Leskelä, M. Bischoff, U. Kaiser, and N. Kaiser, *Appl. Opt.* 47, C271 (2008).
81. T. Pilvi, E. Puukilainen, K. Arstila, M. Leskelä, and M. Ritala, *Chem. Vap. Deposition* 14, 85 (2008).
82. T. Pilvi, E. Puukilainen, F. Munnik, M. Leskel¨a, and M. Ritala, *Chem. Vap. Deposition* 15, 27 (2009).
83. G. Natarajan, P. Maydannik, D. C. Cameron, I. Akopyan and B. V. Novikov *Appl. Phys. Letts.* 97(24), 241905 (2010).

6

Metals

6.1 Introduction

Pure metal thin films such as, aluminum, copper, and tungsten have been used for many years in the production of integrated circuits. Noble metals which are deposited on metal oxides have multiple applications in sensor, catalysts, and microelectronics. In catalysts noble metal nanoparticles with a controlled size and dispersion are desired. In microelectronics smooth, continuous, and conformal thin films are required. Noble metals are needed as nanometer-thin films on complex geometries in a variety of electrodes and nano-capacitors. As examples noble metals could be applied in the positive channel MOS (PMOS) gate electrodes since their vacuum work function is near the conduction band of Si, or they can be applied as ultrathin barrier films with high uniformities for the microproses-sor unit (MPU) interconnect technology.

In the present chapter we will review the thermal and plasma or radical enhanced ALD (definition of REALD and PEALD are described in Chapter 1) processes for pure metals. Thermal ALD has been widely used to deposit many different oxide films. However, deposition of pure metal films is substantially more difficult. As we will see, plasma enhanced processes can bring certain benefits and improvements to the metal film growth and resulting film properties. At the present time, the deposition of pure metals is being actively researched both by thermal and plasma enhanced ALD processes. For more comprehensive information of exiting ALD processes the reader is directed to the recent review articles [1–2].

6.2 Noble Metals

Metal deposition has been generally understood to be only possible via a reduction mechanism. However, Aaltonen *et al.* made important progress in ALD deposition by realizing that oxidizing agents like oxygen, oxygen plasma, or ozone can be used to create noble metal films [3–9]. In the first studies for noble metal depositions it was seen that an O-pulse produces a reactive layer of metal oxide which then oxidizes the ligands of the noble metal precursor during its next pulse. Electrons are released to reduce the metal ion as in the following equation [10].

$$(M^{q+})\,(O^{2-})_{q/2} \rightarrow M^0 + q/2(O^0) \qquad (6.1)$$

A similar release of oxygen has also been reported in a study using Auger electron spectroscopy and in an oxide ALD study [11]. The reactions and mechanisms of combustion chemistry is covered carefully in the work of Elliott [12].

6.2.1 Silver Processes and Applications

Silver, Ag has the highest conductivity of any metals and one of the highest refractive index of all materials. Silver has been considered as precious metal and been used for coins, jewelry, utensils, and mirrors. For long it has been known to be antimicrobial and it has been used in antibiotics and bactericidal products. At present times silver has been used in electrical contacts, conductors, gas sensors, and catalysts in chemical reactions. In the future silver could also replace copper as a conducting material in integrated circuits since it can maintain its low resistivity better than copper in line widths below 100 nm [13].

Silver has not been deposited with ALD before probably due to lack of urgent need in applications but also due to the fact that a proper precursor has been missing. A novel process has been developed by synthesizing a silver precursor and applying a radical-enhanced ALD (REALD) [14]. (2,2-dimethylpropionato)silver(I) triethylphosphine, $Ag(O_2CtBu)(PEt_3)$ has been used as the silver source and hydrogen radicals as reducing agent. The Ag precursor melts at low temperature and has sufficient thermal stability but it suffers from a low decomposition temperature which will make the ALD window narrow. The Ag films were deposited on p-type Si and glass substrates. During the deposition the Ag precursor was evaporated at 125°C and the reaction temperature was 140°C. Growth rate saturation was found and the growth rate was 1.2Å/cycle. The films were found to be conformal possessing low resistivity and relatively low impurity levels. Figure 6.1 presents a SEM image of silver deposited on silicon.

In a recent study plasma enhanced ALD was applied to create Ag films exhibiting plasmonic properties which were varying according to the microstructure of silver film [15]. The investigators found air gaps within PEALD-grown film and concluded the Ag film to be metamaterial and its behavior was close to so called spoof plasmonics where gaps are fabricated in films to create plasmonic resonances.

Figure 6.1 A SEM image of a 60 nm thick silver film deposited on silicon. Image taken at 80000 × magnification [14]. *Reprinted with permission from Wiley-VCH Verlag GmbH Co,© 2007.*

6.2.2 Ruthenium Processes and Applications

First ALD depositions for noble metals were done with ruthenium and platinum [3–5]. A new process for ruthenium ALD films deposited at 275–400°C [4] has been developed. The precursors were bis(cyclopentadienyl)ruthenium ($RuCp_2$) and oxygen. The grown ruthenium films were polycrystalline metallic ruthenium and possessed excellent conformality. The impurity levels were found low in the films. The film resistivity was measured to be 20 μΩ-cm.

In the following study for Ru another precursor was applied [5]. Tris(2,2,6,6-tetramethyl-3,5-heptanedionato)ruthenium and oxygen were used at a reaction temperature of 325–450°C. The films were found to be polycrystalline metallic ruthenium with a good adhesion to Al_2O_3 substrate. The growth rate at 350° was 0.36 Å/ cycle and the films resistivity was 20 μΩ-cm. Impurity levels were measured to be <2.9 at-% of H, <1.9 at-% of C, and <5.5 at-% of O.

Yet another ruthenium precursor was tested. Ruthenium films were deposited with bis(N,N′-di-tert-butylacetamidinato) ruthenium(II)dicarbonyl and oxygen at 300–400°C onto SiO_2(300nm)/Si wafers [16]. The crystallinity was strongly dependent on the O_2 exposure E_o which was determined based on the molecules dosed per ALD cycle, the pumping speed, and the deposition temperature. The best quality films were obtained when the E_o reached its optimum value. The film resistivity was 10 μΩ-cm and roughness was 2 nm for 30 nm thick films. The impurities were reported to be about 0.24–0.30 at-% for oxygen and between 0.25–0.35 at-% for carbon.

Ruthenium has been studied as a potential candidate for the seed layer in electroplated copper interconnects, and electrode material for DRAM capasitors and MOSFETs [1]. Deposition of ruthenium by thermal ALD was shown to produce high purity and low resistivity films. However, there can be several problems with such films. One, may be that very thin ruthenium films will not nucleate properly on all surfaces. In particular, nucleation on some oxides may be inadequate [17]. Another area of concern may be with deposition on metallic diffusion barrier films. Here, it is likely that if oxygen is used as the reducing agent, the barrier metal will be oxidized causing its resistivity to increase [18].

To avoid oxygen incorporation in the film, one approach is to use a PEALD process instead of an oxygen based processeses. There are several PEALD processes that have studied growing ruthenium. The most studied ruthenium precursor is bis(ethylcyclopentadienyl) ruthenium (Ru(EtCp$_2$)) with both NH_3 [18–24]· and H_2/N_2 [25–28] plasmas for precursor reduction.

In the first REALD ruthenium study by Kwon *et al.* [18] the experiments were carried out in a capacitively coupled parallel plate reactor operated at 13.56 MHz and 100 Watts. The deposition temperature was 270°C. A vapor pressure of 180 mtorr was provided by heating the ruthenium precursor to 80°C. A deposition rate of 0.38 Å/sec was observed. The film resistivity was 12 μΩ-cm, and the film density was 12.03 gm/cm^3. No carbon, oxygen or nitrogen impurities were reported.

The question of film nucleation has been addressed in later studies [21–23]. In the study by Kwon *et al.* comparison between the morphology of ruthenium films deposited onto TiN by the PEALD process and by a thermal ALD process using oxygen was made. As already stated in the earlier study [18], it was confirmed that the PEALD film is smoother than the thermal ALD film. As can be seen in Figures 6.2 and 6.3 around 40 ALD cycles were needed before the film growth initiated by thermal ALD, whereas by PEALD the film formation with high nucleation density can be observed already at very early stage of PEALD cycles. Thus PEALD provides less delay in ruthenium nucleation than thermal ALD, at least on TiN [23].

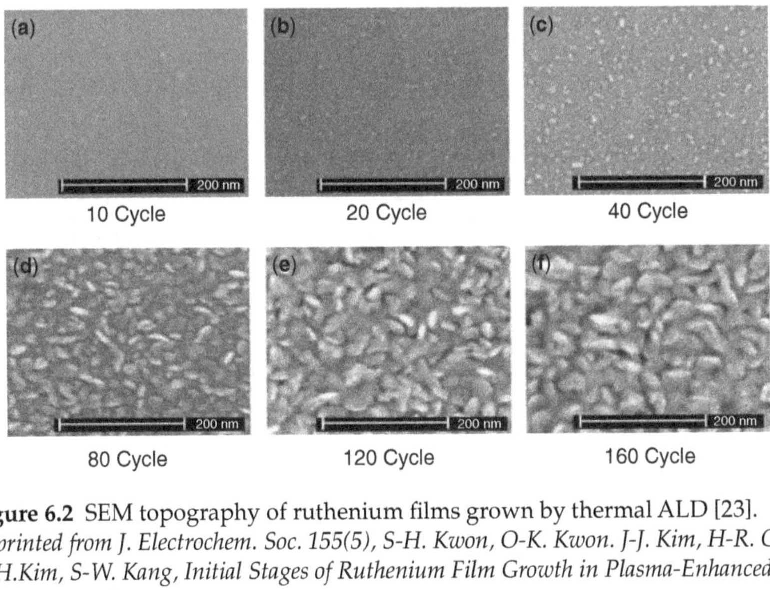

Figure 6.2 SEM topography of ruthenium films grown by thermal ALD [23]. *Reprinted from J. Electrochem. Soc. 155(5), S-H. Kwon, O-K. Kwon. J-J. Kim, H-R. Oh, K-H.Kim, S-W. Kang, Initial Stages of Ruthenium Film Growth in Plasma-Enhanced Electrochemical Society.*

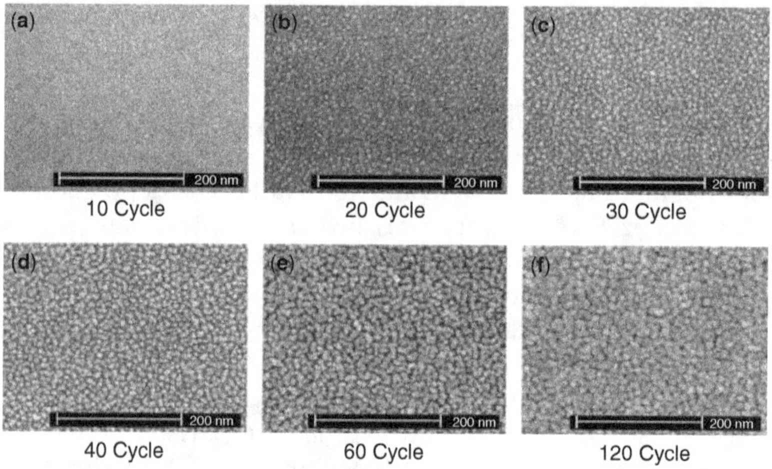

Figure 6.3 SEM topography of ruthenium films grown by PEALD [23]. *Reprinted from J. Electrochem. Soc. 155(5), S-H. Kwon, O-K. Kwon. J-J. Kim, H-R. Oh, K-H. Kim, S-W. Kang, Initial Stages of Ruthenium Film Growth in Plasma-Enhanced Atomic Layer Deposition, H296,, © 2008, Reproduced by permission of ECS – The Electrochemical Society.*

As expected the substrate surface is also one of the factors determining the film nucleation. Same has been observed with PEALD ruthenium, where there is more delay in the film growth on Si than on TiN [21].

Formation of a dense continuous ruthenium film on TiN by nucleation requires about 80 ALD cycles resulting in a 2.7 nm thick film. After this transient region the film growth is linear. It is notable that the transient region depends on the length of $Ru(EtCp_2)$ pulse as well. The full film coverage with 25 s precursor pulse was achieved after 80 cycles whereas more than 150 cycles were required with 7 s precursor pulse [23]. The same transient region in ruthenium growth was found in the following experiments, which were done on TiN and silicon surfaces at deposition temperature of 350°C. The full coverage of ruthenium on TiN surface was reached after 50 cycles and on silicon after 80 cycles [21]. The delay in film nucleation has been a common observation since the first ruthenium ALD process published [3]. In the case of $Ru(EtCp_2)$ the delay in nucleation has been attributed to a low adsorption probability of the precursor on the native surface [23]. New precursor such as (2,4-dimethylpentadienyl)(ethylcyclopentadienyl) ruthenium (DER) may provide a way to decrease this delay in film nucleation period [29].

The film performance in general is determined by its structure. This is especially the case in the potential application for ruthenium, where the desired film thickness is in the range of few nanometers. The structure should rapidly achieve a certain orientation when nucleation occurs quickly. Here the REALD process can also provide advantages over the oxygen based thermal ALD [21].

6.2.3 Platinum and Palladium Processes and Applications

Platinum was grown at 300°C using (methylcyclopentadienyl)trimethylplatinum (MeCpPtMe3) and oxygen as precursors [6]. The produced films were uniform, had low resistivity, and low impuritity levels. X-ray diffractions showed that the films were strongly oriented in <111> direction. The growth was found to be linear and the rate was 0.45 Å/cycle.

Palladium and platinum were deposited onto ALD grown Al_2O_3, ZrO_2 or TiO_2 films [30]. The ALD processes for 10–100 nm metal oxide films were TMA/H_2O, Bis(cyclopentadienyl)dimethyl Zirconium, ($ZrCp_2Me_2$)/H_2O and $TiCl_4$/H_2O. Palladium films were deposited with palladium (II) hexafluoroacetylacetonate ($Pd(hfac)_2$) and formalin (HCOH) at 200°C. The precursor was evaporated in a bubbler at 50°C. Platinum was deposited with trimethyl(methylcy clopentadienyl) platinum(IV) ($Pt(MeCp)Me_3$) and high purity oxygen (99.999%) at a temperature of 300°. The precursor was kept in a bubbler at 30°C. The film growth was followed using *situ* QCM and QMS measurements. It was discovered that the nucleation of Pd and Pt was slow on Al_2O_3 substrates most probably because the growth took place by a coalescence of islands. The Pd nucleation was hastened by increasing the pulsing time. Changing the reducing agent after the first pulse of HCOH into H_2 further helped the islands to agglomerate faster. Pt film growth depended strongly on the metal oxide substrate. On ZrO_2 the films grew nearly continuously whereas on TiO_2 isolated Pt nanoparticles were formed. On Al_2O_3 Pt grew as high density of nanoparticles. The results were very interesting concerning the various applications in catalysts or in microelectronics. In the Figure 6.4 SEM images of AFM tips are presented. It can be seen how the Pt grows continuously on the ZrO_2 layer creating a smoother surface than on the commercial Au-coated AFM tip.

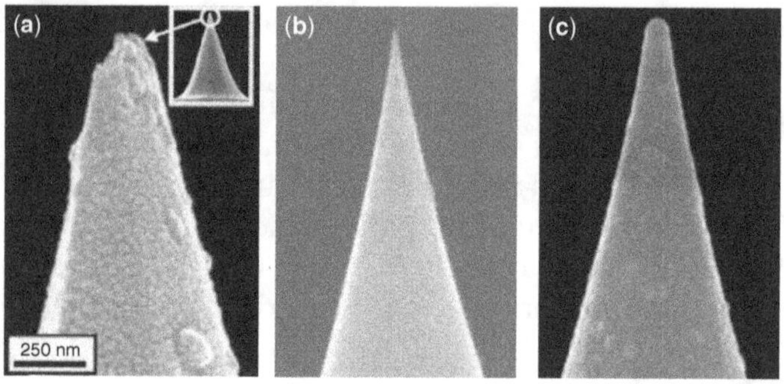

Figure 6.4 SEM images of (a) commercial Au-coated AFM tip, (b) uncoated Si AFM tip, and (c) the same Si AFM tip following 5 nm ALD ZrO_2 and 30 nm ALD Pt [30]. *Reprinted from ECS Transactions, 3, J.W. Elam, A.V. Zinovev, M.J. Pellin, D.J. Comstock, M.C. Hersam, Nucleation and growth of noble metals on oxide surfaces using atomic layer deposition, 271, © 2007, Reproduced by permission of ECS – The Electrochemical Society.*

Deposition of palladium films have been reported using palladium(II)hexafluoroacetylacetonate as the precursor and atomic hydrogen [31] or mixture of hydrogen and nitrogen [32] as the reducing agent. In particular, films were grown on oxide and polymer surfaces, where thermal ALD had not been successful. The experimental setup was a downstream plasma arrangement, similar to that used to deposit titanium and tantalum [33–36] (see Figure 6.7).

In the first study by this group the film growth was carried out at 80°C for 150 cycles with a 65 watt solenoidal source. One substrate used had a thin film of tungsten along with its associated oxide layer. The other was a silicon wafer with a native oxide coating that had been dipped in toluene to approximate the surface of a low dielectric constant electrical insulator. The growth rate on the tungsten sample was ~ 0.25 Å/cycle, and there was little evidence of an incubation period. On the other hand, the initial growth rate on the silicon sample was 0.05 Å/cycle, but after 100 cycles stabilized at 0.22 Å/cycle. Clearly there is some complex surface chemistry going on when the deposition surface is not yet coated with palladium [31].

None of the films displayed any fluorine contamination, but they did contain 6–10% carbon. Surface morphology was examined with AFM, and showed that the film deposited on tungsten replicated the original surface, while the film on silicon dioxide was quite rough. On a silicon substrate with a starting roughness of 5.32 Å, a deposited 16.6 Å thick Pd film was measured to have a roughness of 21.25 Å. Clearly, for this example, these films did not nucleate in a uniform manner. Finally, it was interesting to observe that although this film was quite rough, it was still amorphous [31].

The following study by this group concerned PEALD palladium on a polymer (poly(*p*-xylylene), PPX) substrate. Again the film growth was carried out at 80°C for 100 PEALD cycles with a 50 W H_2:N_2 plasma. The PPX substrate was completely etched by a pure H_2 plasma. By mixing hydrogen with nitrogen in a ratio of 1:1.5 it was able to avoid the etching and deposite palladium on PPX. Contrary to previous study with pure H_2 plasma films in this study showed low fluorine and slightly higher carbon contamination indicating incomplete ligand removal during the plasma step of ALD cycle. The growth rate on the air exposed silicon substrate was 0.38 Å/cycle and on PPX polymer substrate 0.25 Å/cycle. The higher growth rate with H_2:N_2 plasma compared to H_2 plasma was attributed to increase in chemisorption sites for the precursor. The lower growth rate on PPX was attributed to the fact that the adsorption of palladium monolayer and the film nucleation is dependent on the active surface sites created by plasma during the ALD cycle [32]. This will obviously result in a slower film growth regime until the complete monolayer of palladium is formed on the polymer surface.

6.2.4 Rhrodium Processes and Applications

Rhodium was deposited at 300°C using rhodium(III)acetylacetonate and oxygen as precursors [37]. CVD grown HfO_2 and thermally grown SiO_2 were used as substrates. The growth was found self-limiting and the growth rate was 1.9–2.1 Å/cycle which is slightly higher compared to Rh-process conducted by Aaltonen *et al* [8]. The effective work function was measured to be 5.43 and 5.25 on SiO_2 and HfO_2, respectively. It is very common that in the annealing process oxygen is needed to prevent oxygen vacancy formation at the dielectric/metal interface. Consequently the creation of oxygen vacancies will change the electrical properties. The authors concluded that in the Rh-process the residual precursor oxygen may prevent the formation of oxygen vacancies hence resulting in the high work function values.

6.2.5 Iridium Processes and Applications

Currently a Ta/TaN barrier structure is used in Cu interconnections in microprocessor units (MPUs). ALD grown nobel metals are a potential replacement for Ta/TaN since they allow direct electroplating of a Cu layer on the barrier without the need for a seed layer. Lim *et al* [38] grew iridium films onto silicon using (ethylcyclopentadienyl)(1,5-cyclooctadiene) iridium [Ir(EtCp)(COD)] and oxygen as precursors within a temperature range of 240–420°C. Argon was used as a purging gas. The ALD conditions were found at 330–360°C where a smooth and uniform Ir film was grown. Figure 6.5 presents cross-sectional TEM images of the growth of Ir film at various cycles.

The TEM images in Figure 6.5 show the Ir film growth at 100, 120, 170, 200, and 300 cycles. The nucleation is seen as faint dots at 100 cycles. Compared to Ru studies the incubation time is clearly longer [39]. Between 100 and 170 cycles an island growth is noticeable. At 200 cycles the islands start to agglomerate but do not yet form a continuous film. It seems to require 300 cycles to obtain a continuous Ir film. The thickness of this film was found to be 10 nm. Based on XRD spectrum the film was found to be polycrystalline with a main orientation of <111>. The investigators provided an illustrated growth mechanism for Ir film growth which is presented in Figure 6.6.

In another study Iridium was grown to serve as a seed layer for Cu in a study where (ethylcyclopentadienyl)(1,5-cyclooctadiene) iridium [Ir(EtCp)(COD)] and NH_3 were used as precursors at

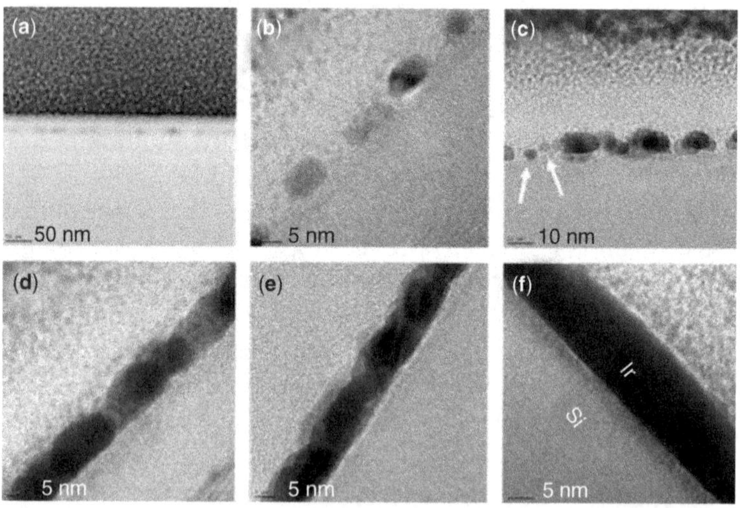

Figure 6.5 Cross-sectional TEM images of Ir thin films deposited on silicon substrate at 330°C with (a) 100, (b) 120, (c) 170, (d) 200, (e) 250, and (f) 300 cycles [38]. *Reprinted with permission from Wiley-VCH Verlag GmbH & Co, © 2011.*

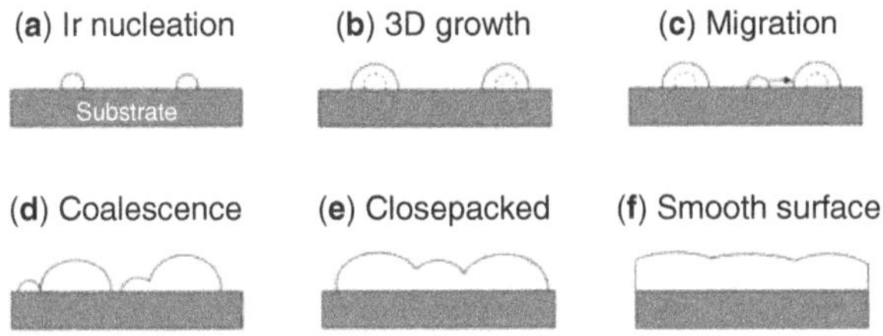

Figure 6.6 Illustration for growth mechanism of ALD Ir films in each cycle (a) 100, (b) 120, (c) 170, (d) 200, (e) 250, and (f) 300 cycles [38]. *Reprinted with permission from Wiley-VCH Verlag GmbH & Co, © 2011.*

290°C [40]. The growth rate was found to be 0.55 Å/cycle and the film resistivity was 8.3 μΩ-cm.

6.3 Titanium

Growth of Ti from a titanium halide and hydrogen with a CVD process is known to require temperatures greater than 1500°C before acceptable deposition rates are achieved. Therefore, the only practical method of depositing such a metallic film is to use a much more energetic reducing agent. The obvious choice would be H atoms.

Pure titanium films have been deposited by REALD from $TiCl_4$ using a hydrogen glow discharge to produce hydrogen atoms [33]. The REALD reactor configuration used is shown in Figure 6.7.

In much the same way as the system described in Figure 2.8, the plasma is generated in a quartz tube surrounded by a solenoidal coil powered by a 13.56 MHz RF generator at powers between 300 and 1200 watts. In this case, however, hydrogen rather than helium flows through the tube. Because the substrate does not sit in the glow discharge this arrangement is referred to as a "downstream" radical source. The recombination rate of electrons and ions is extremely rapid, so they do not survive much beyond the end of the coil. However, the atomic radicals recombine more slowly into molecular species, so they are available to reduce the $TiCl_4$ adsorbed onto the substrate surface. Some of the later papers that we will be reviewing, describe deposition in reactors that create their glow discharges in a capacitor configuration, where the substrate sits in the glow discharge.

Depositions were carried out in a 6–8 liter volume stainless steel vacuum chamber which was pumped by a 500 l/s turbopump, and

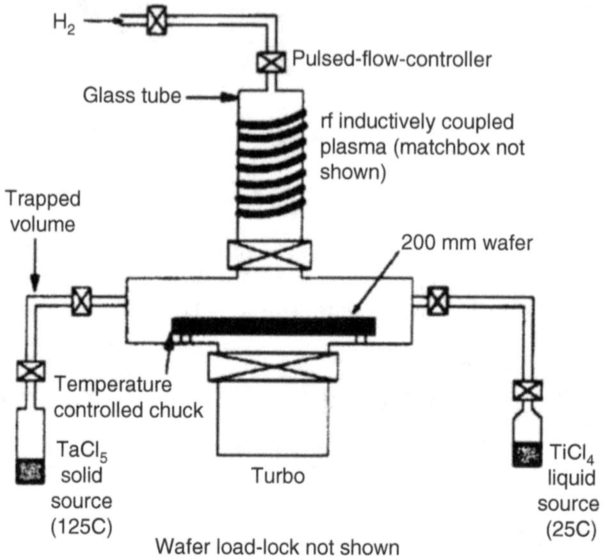

Figure 6.7 Schematic of REALD reactor system [33]. *Reprinted with permission from S. M. Rossnagel, A. Sherman, F. Turner, J. Vac. Sci. Technol. B18 © 2000, American Vacuum Society.*

had a base pressure of $< 10^{-07}$ torr. The source was a quartz tube filled with liquid $TiCl_4$, and held at room temperature. Its vapor pressure was on the order of 10 torr. This was sufficient to enable dosing (over 100,000 Langmuirs) of the substrate with a monolayer of $TiCl_4$ with a short exposure time (~ 1 sec). The silicon substrate was held at a temperature of 250°C, and films ranging from 20–150Å were deposited.

Analysis of the pure Ti films, on bare silicon, were complicated by their rapid oxidation when exposed to air for a few seconds. This rapid oxidation of REALD Ti films, compared to sputtered films of comparable thickness, suggests films of lower density. RBS analysis of the oxidized films showed stoichiometric TiO_2 with only one percent chlorine contamination. Additional experiments where Ti films were deposited onto Cu coated wafers at 25°C, yielded more stable metallic films. In fact, when copper was sputtered on top of REALD titanium, *in situ*, no oxygen contamination was found in the titanium after air exposure, by RBS analysis [34].

A simplified version of the chemistry being explored, is shown below

$$TiCl_4(g) + 4H \rightarrow Ti(s) + 4HCl \tag{6.2}$$

Although this process follows this overall equation, many questions as to the details of the process remain. For example, when $TiCl_4$ adsorbs to a surface does it chemisorb intact or does it dissociate after adsorbing. Evidence seems to support dissociative chemisorption [34], where the surface species are $TiCl_x$ with x on the order of 2.

Finally, this process appears to have a non-linear growth rate for the initial cycles, depending on the starting substrate material. For one class of substrate materials (i.e. platinum, aluminum, gold and copper), the initial growth rate per cycle is more rapid than the linear rate observed after perhaps 10 cycles. The opposite behavior is seen for deposition on silicon and amorphous carbon. Data generated with a quartz crystal microbalance are shown in Figure 6.8.

The initial rapid growth on Pt may be due to the profusion of sites on which the $TiCl_4$ can dissociate, at the start of deposition. After these sites are covered over by deposited titanium, the deposition rate becomes somewhat slower, but very linear, as we would expect for an ALD process. However, when initial deposition is on silicon, we see the well known phenomena that hydrogen atoms will etch silicon. Then, once all of the silicon is coated with Ti, the growth becomes linear.

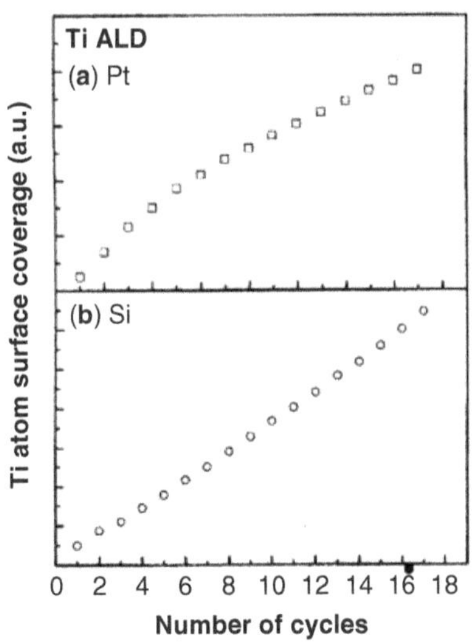

Figure 6.8 Thickness of Ti growth vs growth cycles, for two substrate materials. Room temperature deposition with 60 second exposures [34]. *Reprinted with permission from H. Kim, S. M. Rossnagel, J. Vac. Sci. Technol. A20, 802 © 2002, American Vacuum Society.*

6.4 Tantalum

As is true for Ti, no precursor has been found that will allow deposition of Ta by thermal ALD. However, it can be deposited from the halides by REALD using hydrogen atoms as the reducing agent. The particular halide precursor that has been evaluated recently is $TaCl_5$, which is a thermally stable solid powder. When heated to 100°C, sufficient vapor pressure is available to enable a monolayer of $TaCl_5$ to form quickly on a substrate. The same REALD reactor, as shown in Figure 6.7, was used. As before, the complex reaction mechanism for this reaction can be expressed simply as

$$TaCl_5(ad) + 5H(g) \rightarrow Ta(s) + 5HCl \tag{6.3}$$

Experiments to determine film characteristics as a function of deposition parameters were carried out on silicon wafers [35]. Good quality films were reportedly grown at temperatures as low as 150°C. For a deposition temperature of 250°C, the deposition rate was found to saturate, for exposure times above 1 second, at 0.08 Å/sec. The only impurities found in these films were small amounts of chlorine (< 10%), and hydrogen (~10–20%). The chlorine content and the film resistivity were closely correlated. Both were substantially reduced by extended exposure to the hydrogen plasma, as seen in Figure 6.9. The value of resistivity found for hydrogen plasma exposure > 4 sec was 150–180 μΩ-cm.

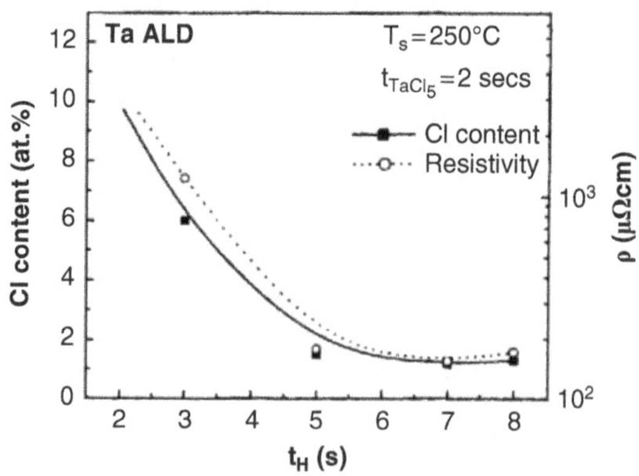

Figure 6.9 Chlorine content and resistivity at 250°C as a function of atomic H exposure time [35]. *Reprinted from Thin Solid Films, 441, H Kim,S.M Rossnagel, Plasma-enhanced atomic layer deposition of tantalum thin films: the growth and film properties, 311 © 2003 with permission from Elsevier.*

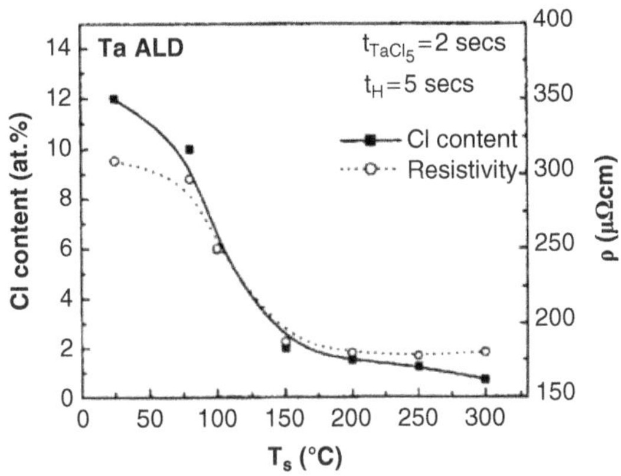

Figure 6.10 Chlorine content and resistivity with hydrogen exposure for 5 seconds as a function of deposition temperature [35]. *Reprinted from Thin Solid Films, 441, H Kim,S.M Rossnagel, Plasma-enhanced atomic layer deposition of tantalum thin films: the growth and film properties, 311 © 2003 with permission from Elsevier.*

This value of resistivity corresponds closely to values for the β Ta phase. In addition, the chlorine content and the film resistivity could be reduced by increasing the deposition temperature, as can be seen in Figure 6.10. Here it can be seen that even at deposition temperatures as low as 150°C, the Ta film resistivity was below 200 μΩ-cm and the chlorine content was below 2%.

Making use of XRD spectra and TEM images of these films it was demonstrated that the, as grown, films are amorphous. It was also found by RBS analysis that annealing films deposited on SiO_2 at 250°C for 30 minutes reduced the chlorine and hydrogen concentrations substantially, as shown in Figure 6.11.

Up to annealing temperatures of 750°C, all films remained substantially amorphous. Above this annealing temperature, the increase in the silicon signal suggests that inter diffusion of tantalum and silicon had occurred, possibly suggesting some crystallization of the tantalum. Also, at this annealing temperature, these measurements showed that the chlorine content had been reduced to about 1%.

One of the more significant applications of thin pure tantalum films is their use as diffusion barriers. Since copper has begun to replace aluminum as the electrical conductor in the newest integrated circuits (due to its lower electrical resistance), a diffusion barrier between the copper and silicon oxide insulators becomes essential. This is because copper can diffuse very rapidly in such insulators, and influence the circuits' performance. Traditionally, sputtered tantalum has been used as the diffusion barrier in this

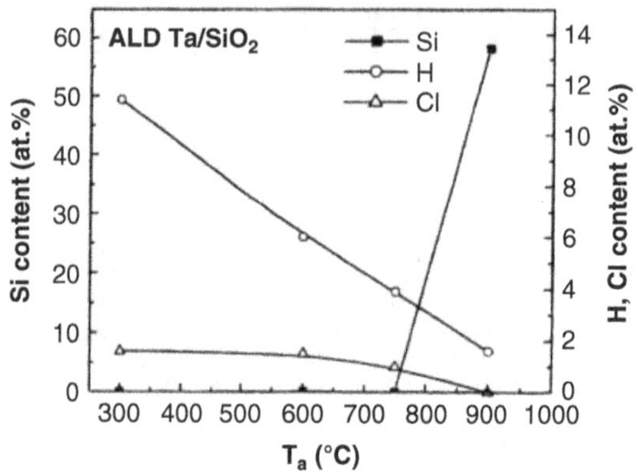

Figure 6.11 RBS measurements of Cl, H, and Si in 250°C films as a function of annealing temperature Ta [35]. *Reprinted from Thin Solid Films, 441, H Kim, S.M Rossnagel, Plasma-enhanced atomic layer deposition of tantalum thin films: the growth and film properties, 311 © 2003 with permission from Elsevier.*

application. Because sputtering is a line of sight physical deposition method, coating high aspect ratio holes is difficult. Any ALD technique will do better in terms of conformal coatings, and the same is true for the REALD of tantalum.

Thin tantalum films deposited at 400°C by REALD were compared to similar films deposited by sputtering (PVD) [36]. Films were on the order of 28–140Å for the ALD films, and 50–250Å for the sputtered films, and were all deposited onto silicon wafers. A 2000Å thick copper film was sputtered onto each tantalum sample to provide a test for diffusion barrier effectiveness. Barrier failure was defined by a number of techniques, one of which determined the presence of copper silicide in the Cu/Ta/Si stack.

REALD films consistently survived to higher annealing temperatures than the PVD films. Apparently, the improved performance of the REALD films was due to their microstructure. Compared to the PVD films, they remained amorphous or were microcrystalline up to the point of failure. Also, their advantage increased as the films became thinner. For the thicker films they had a 70°C increment, but for films thinner than 50Å it increased to 100°C.

6.5 Aluminum

Although there have been many attempts to grow Al thin films by CVD, as an alternative to sputtering, they all suffered from two defects. First, it proved very difficult to dope such films with copper to improve electromigration resistance. Secondly, the films grown were quite rough as compared to sputtered films, so that it was difficult to control lithographic patterning [41]. Attempts to grow such films by thermal ALD have not been reported.

Using TMA as a precursor, hydrogen plasma can be used to readily reduce it to pure aluminum [42]. Experiments were carried out in a capacitively coupled reactor operated at 13.56 MHz, and at 200 watts. The wafer temperature was held at 250°C. As long as the TMA pulse time was greater than 3 seconds, and the hydrogen plasma pulse was longer than 5 seconds, deposition rate did not depend on pulse times. Unfortunately, no information was provided in this reference as to behavior at other deposition temperatures, so it is not clear what kind of a temperature window exists. Nonetheless, very smooth and conformal films were deposited, as can be seen in Figure 6.12.

The value of the electrical resistivity of a 1000Å thick film was reported to be ~ 10 μΩ-cm, which is quite high compared to the bulk value for aluminum of 2.65 μΩ-cm. Part of the reason for this, may be the fact that the films were reported to contain carbon (~ 3%), and hydrogen (~ 4%) impurities. It is also likely that the aluminum was contaminated with oxygen, but this impurity was not mentioned.

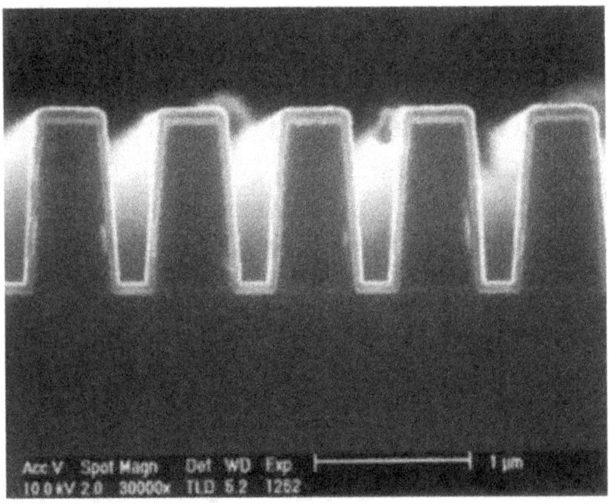

Figure 6.12 Conformality of 500Å Al film on 5:1 aspect ratio trenches [42]. *Reprinted from Electrochem. Solid State Lett. 5, Y. J. Lee, S-W. Kang, C91, © 2002, Reproduced by permission of ECS – The Electrochemical Society.*

Finally, if such films were to be used as replacement for sputtered aluminum, it would be necessary to dope them with silicon and copper. Presumably, this should be possible with other ALD processes.

6.6 Copper

Copper has been studied a lot with ALD whereas iron, cobalt, and nickel have been explored relatively little. The interest in copper thin films has been created by the development of copper rather than aluminum based integrated circuits. The first demonstration that copper could be deposited by this technique carried out the process using CuCl as the copper precursor, and zinc vapor as the reducing agent [43]. As in the previous discussion, any attempt to use hydrogen as the reducing agent would have required too high a deposition temperature to be practical.

Aside from the practical problems associated with the high temperatures needed to establish sufficient vapor pressure from these two precursors (390°C for CuCl, 435°C for zinc vapor), the other concern would be the dissolution of zinc into the growing copper (brass formation). A flow reactor operating at 10 torr was used, and depositions were carried out from 440–500°C. Substrates were 5×5 cm glass coated with aluminum oxide. It was possible to grow copper films by alternating exposures to the precursors, and growth rates higher than ~ 10 Å/cycle were found. Data were reported only for 500°C depositions, so it is not clear if there is an acceptable temperature window for this process. In addition, the film growth rate increased substantially with the length of the precursor pulses. Based on these facts it would appear that there was a considerable CVD component to these depositions.

The films were contaminated with about 3% zinc, even with the shortest pulse possible (0.5 sec). Also, they nucleated readily onto Al_2O_3 coated glass, but were weakly adherent.

A second study, done at about the same time, deposited copper from an organometallic precursor using hydrogen as the reducing agent [44].

The precursor was $Cu(thd)_2$, Cu(II)-2,2,6,6-tetramethyl-3, 5-heptandionate. It was evaporated at 120°C, and depositions were carried out over the temperature range from 150 to 350°C. Under these conditions, the only substrate that allowed any deposition was a glass substrate with a thin coating of a mixture of platinum and palladium. A temperature window from 200–275°C, was reported, with a deposition rate of 0.35 Å/cycle. Film resistivity was at least five times higher than bulk copper. Based on the reported results, primary interest in this process will be in applications for selective growth of copper.

In another investigation of the thermal ALD of copper, a different organometallic compound was used [45]. It was $Cu(hfac)_2 \cdot XH_2O$ or copper(II)- 1,1,1,5,5,5-hexafluoroacetylacetonate hydrate. In fact, this precursor has been studied in many cases of CVD of copper. Reducing agents that were tried were methanol, ethanol, and formalin. Films were grown at temperatures as low as 230°C, but the only experimental results reported were carried out at 300°C. Depositions were done on 60Å thick barrier layers of Ta, TaN, and TiN as well as glass. Adhesion was reported to be good on TiN and TaN, but not as good on glass or Ta. Smooth bright copper colored films were grown using formalin, and the resistivity was 4.25 μΩ-cm for a thin 200Å film and 1.78 μΩ-cm for a thicker 1200Å film. The authors only report a trace of carbon impurity and less than 2% of oxygen. An SEM of the thinner film coating high aspect ratio trenches showed the conformality to be excellent.

Although these results are certainly promising, it would also be of interest to know more about a temperature window, as well as whether or not deposition rate depends on the level of exposure of the substrate to each precursor.

Another approach to the deposition of copper thin films by ALD, is to use energetic radicals to reduce the copper precursor, rather than relying on substrate temperature to induce the reaction. REALD will not only provide a way to decrease to deposition temperature but also a way to decrease the impurity level in the film due to a more efficient reduction of copper precursor. Initially REALD grown copper film was demonstrated by Niskanen *et al.* [46] by using copper(II)acetylacetonate [$Cu(acac)_2$] and atomic hydrogen in a downstream REALD reactor. Later the same process was used to study the usage of the ALD copper as a seed layer in electroplated interconnects. ALD copper layer as a seed layer for copper electroplating may provide significant advantages in Cu interconnects used in nanoelectronic applications due to ALD film conformality, thickness uniformity and consequently improvement in electrical performance [47].

In Niskanen's et al. work the deposition was carried out at temperature of 140°C. A growth rate of 0.18Å/cycle was achieved with a 300 watt microwave power source. Saturated growth was confirmed with a 1 second pulse of the copper precursor, and a 5 second pulse of hydrogen atoms [46].

The copper films grown contained 11 at. % oxygen, 2 at. % hydrogen, and 1 at. % carbon. The film resistivity was measured for 250Å thick films, and was found to be ~ 15 μΩ-cm, which is substantially higher than the value for bulk copper. This elevated resistivity was again likely due to oxygen contaminant in the film.

The films were observed to be polycrystalline, with a roughness of 2.5–3Å for a 250Å thick film. They also had good adhesion to all substrate surfaces, and as expected showed excellent conformality. This improved adhesion is a significant advantage over thermal ALD copper films, when used as a seed layer for copper plating.

In another study the deposition was carried out at temperature of 85°C in order to minimize the thermally induced agglomeration and surface roughness of copper film. The study was done with direct plasma configuration with a plasma mesh parallel to the wafer. A growth rate was slightly higher than in earlier work by Niskanen *et al.*, being 0.20 Å/cycle. The film composition measured by Auger electron spectroscopy showed more that 95% copper purity in the film with the oxygen and carbon composition at the measurement noise level. The resistivity of a 12 nm thick copper film deposited on a SiO_2 was 38.8 μΩ-cm whereas the resistivity of 40 nm thick layer was measured to be 7.6 μΩ-cm [47].

A more recent study provides information for the PEALD temperature window by discussing the behavior of film growth characteristics and film properties subject to deposition temperature between 100 and 350°C. Although, it must be noted that the study was done using a different copper precursor, Bis(1-dimethylamino-2-methyl-2-butoxy)copper (MABOC), than in earlier studies. Counter reactant was provided by H_2 plasma with radio frequency configuration. The film growth between the deposition temperature of 100 and 180°C was found to follow the self-limiting growth mode with the growth rate of around 0.65 Å/cycle. Above 180°C the growth rate increased dramatically due to the decomposition of copper precursor leading to CVD film growth. The film composition measured by X-ray photoelectron spectroscopy showed carbon and nitrogen composition of around 2% in the film. The resistivity of a 10 nm copper film decreased slightly with increasing deposition temperature and reached the lowest value of 5.2 μΩ-cm at the deposition temperature of 150°C [48].

6.7 Other Transition Metals

Molybdenum and tungsten have both been deposited from halides. Both of these high temperature melting point refractory metals have many potential applications as extremely thin films. One study attempted deposition of molybdenum by a thermal ALD process from $MoCl_5$ with zinc vapor as a reducing agent [49]. The obvious choice for a reducing agent would have been hydrogen, but for a thermal process the deposition temperature would have been quite high. Aside from the practical difficulty of introducing zinc vapor from an evaporator operated at 390°C, growth rate and zinc concentration in the molybdenum film varied with the purge length after zinc vapor pulses, as well as the deposition temperature. The maximum growth rate in these experiments was found at 420°C, with a 0.5 sec purge, to be 0.8 Å/cycle. However, under these conditions the zinc concentration in the film was 2.5%. It was only when the purge was increased to 3 sec that the zinc level fell to 0.5%. Unfortunately, this change reduced the deposition rate by a factor of two. Also, the chlorine concentration was ~ 1–2%, regardless of deposition temperature or purge length.

The molybdenum films grown were polycrystalline, and as long as the zinc concentration was kept at no more than 1%, film resistivity was ~ 15 µΩ-cm as compared to a bulk value of 5.7 µΩ-cm. Also, the films were adherent to soda lime glass and Al_2O_3.

Tungsten W is hard, refractory metal that is relatively inert, and is used for electrical lighting and via filling for microelectronic circuits. Tungsten films can be grown from WF_6 by using Si_2H_6 as a reducing agent. With these precursors a reasonably wide temperature window was found between 150 and 325°C [50]. Also, pure tungsten films were reportedly grown with no reported contamination from fluorine or silicon. As well, the films were either amorphous or composed of very small grains, and so were quite smooth. The disadvantage of this film morphology is apparently, that the film resistivity is high (122 µΩ-cm) compared to bulk tungsten (5 µΩ-cm).

The tungsten films were discovered to grow differently on different substrates [51]. When such films are grown on SiO_2, they require about 10 cycles of deposition before growth proceeds at a steady rate of 3.5 Å/cycle. Tungsten grown on Al_2O_3 begins growing immediately at a rate of 3.5 Å/cycle. This is one illustration of an incubation period before uniform film growth rate can occur. This effect has significance in that it is found that if film growth does not initiate immediately, some island growth may occur to begin the deposition. This can lead to rougher films when the islands coalesce to form a continuous layer.

Another set of organometallic precursors have been developed for use in deposition of transition metals [52] that would include copper as well as iron, cobalt and nickel. The cobalt version is bis(N,N′-diisopropylacetamidinato)cobalt(II). For thermal ALD of these metals, the authors report that each of these precursors is sufficiently volatile at temperatures below 130°C. Experimental results reported demonstrate self limiting behavior for a 250°C copper process, and a 300°C cobalt process. Deposition rates are in line with reported values in the earlier processes described.

Fe, Co, Cu, and Ni films are reported to be polycrystalline, shiny metallic, and specular. They are all also described as electrically conductive, but the only measured data reported was a resistivity value of 46 μΩ-cm for a 400Å thick cobalt film. A cross-sectional SEM is presented to confirm that these films can exhibit exceptional conformality.

However, there does not appear to be any temperature window for either of the processes reported. Therefore, if any of these films are to be grown commercially, it will be a challenge for the reactor designer to maintain very precise temperature control over large substrates, in order to ensure the excellent thickness control expected of ALD. This will be especially true for the very thin films being described.

References

1. V. Miikkulainen, M. Leskelä, M. Ritala, R. L. Puurunen, *Appl. Phys. Rev*, 113, 021301 (2013).
2. H. B. Profijt, S. E. Potts, M. C. M. van de Sanden, W. M. M. Kessels, J. *Vac. Sci. Technol. A* 29(5), 050801 (2011).
3. T. Aaltonen, P. Alen, M. Ritala, M. Leskelä, *Chem. Vap. Deposition* 9, 45 (2003).
4. T. Aaltonen, M. Ritala, K. Arstila, J. Keinonen, M. Leskelä, *Chem. Vap. Deposition* 10, 215 (2004).
5. H.T. Wang, R.G. Gordon, R. Alvis, R.M. Ulfig, *Chem. Vap. Deposition* 15, 312 (2009).
6. T. Aaltonen, M. Ritala, T. Sajavaara, J. Keinonen, M. Leskelä, *Chem. Mater.* 15, 1924 (2003).

7. J. Hämäläinen, F. Munnik, M. Ritala, M. Leskelä, *Chem. Mater.* 20, 6840 (2008).
8. T. Aaltonen, M. Ritala, M. Leskelä, *Electrochem. Solid-State Lett.* 8, C99 (2005).
9. T. Aaltonen, M. Ritala, V. Sammelselg, M. Leskelä, *J. Electrochem. Soc.* 151, G489 (2004).
10. T. Aaltonen, A. Rahtu, M. Ritala, M. Leskelä, *Electrochem. Solid-State Lett.* 6, C130 (2003).
11. K. J. Park, D. B. Terry, S. M. Stewart, G. N. Parsons, *Langmuir* 23 (2007).
12. S.D. Elliott, *Langmuir* 26, 9179 (2010).
13. L. Gao, P. Harter, C. Linsmeier, A. Wiltner, R. Emling, D. Schmitt-Landsiedel, *Microelectron. Eng.* 82, 296 (2005).
14. A. Niskanen, T. Hatanpää, M. Leskelä, M. Ritala, *Chem. Vap. Deposition* 13, 408 (2007).
15. S.M. Spokes, D.J. Glembocki, E. Cleveland, J.D. Caldwell, E. Foos, J. Niinistö, M. Ritala, *Appl. Phys. Lett.* 100, 53106 (2012).
16. H. Wang, R.G. Gordon, R. Alvis, R.M. Ulfig, *Chem. Vap. Deposition* 15, 312 (2009).
17. H. Kim and S. M. Rossnagel, *US Patent Application* 20050118807, June 2, 2005.
18. O. K. Kwon, S. H. Kwon, H. S. Park and S. W. Kang, *Electrochem. Solid-State Lett.* **7**, C46 (2004).
19. M. G. Ko, W. S. Kim, S. K. Park, H. D. Kim, J. W. Park, *J. Korean Phys. Soc.* 53, 2123 (2009).
20. S. S. Yim, D. J. Lee, K.S. Kim, M. S. Lee, S. H. Kim, K. B. Kim, *Electrochem. Solid-State Lett.* 11, K89, (2008).
21. J. Musschoot, Q. Xie, D. Deduytsche, K. De Keyser, D. Longrie, J. Haemers, S. Van den Berghe, R. L. Van Meirhaeghe, J. D'Haen, C. Detavernier, *Microelectron. Eng.* 87, 1879 (2010).
22. S. J. Park, W. H. Kim, H. B. R. Lee, W. J. Maeng, H. Kim, *Microelectron. Eng.* 85, 39 (2008).
23. S.-H. Kwon, O.- K. Kwon, J. –H. Kim, H. –R. Oh, K. –H. Kim, S. –W. Kang, *J. Electrochem. Soc.* 155(5), H296 (2008).
24. Q. Xie, Y. L. Jiang, J. Musschoot, D. Deduytsche, C. Detavernier, R. L. Van Meirhaeghe, S. Van den Berghe, G. P. Ru, B. Z. Li, X. P. Qu, *Thin Solid Films* 517, 4689 (2009).
25. S. W. Kim, S. H. Kwon, S. J. Jeong, S. W. Kang, *J. Electrochem. Soc.* 155, H885 (2008).
26. S. H. Kwon. O. K. Kwon, J. S. Min, S. W. Kang, J. *Electrochem. Soc.* 153, G578 (2006).
27. S. J. Jeong, D. I. Kim, S. O. Kim, T. H. Han, J. D. Kwon, J. S. Park, S. H. Kwon, J. *Nanosci. Nanotechno*l. 11, 671 (2011).
28. C. C. Chang, F. M. Pan, *J. Electrochem. Soc.* 158, G97 (2011).

29. S. K. Kim, S. Y. Lee, S. W. Lee, G. W. Hwang, C. S. Hwang, J. W. Lee, J. Jeong, *J. Electrochem. Soc.* 154, D95 (2007).
30. J.W. Elam, A.V. Zinovev, M.J. Pellin, D.J. Comstock, M.C. Hersam, *ECS Transactions* 3, 271 (2007).
31. G. A. ten Eyck, J. J. Senkevich, F. Tang, D. Liu, S. Pimanpang, T. Karaback, G. C. Wang, T. M. Lu, C. Jezewski and W. A. Lanford, *Chem. Vap. Dep.* 11, 60 (2005).
32. G. A. Ten Eyck, S. Pimanpang, J. S. Juneja, H. Bakhru, T-M. Lu, G-C. Wang, *Chem. Vap. Deposition* 13, 307 (2007).
33. S. M. Rossnagel, A. Sherman, F. Turner, J. *Vac. Sci. Technol. B*18, 2016 (2000).
34. H. Kim and S. M. Rossnagel, J. *Vac. Sci. Technol. A20*, 802 (2002).
35. H. Kim and S. M. Rossnagel, *Thin Solid Films* 441, 311 (2003).
36. H. Kim, C. Cabral, Jr., C. Lavoie and S. M. Rossnagel, *J. Vac. Sci. Technol.* B20, 1321 (2002).
37. K.J. Park, G.N. Parsons *Appl. Phys. Lett.* 89, 043111 (2006).
38. Y.H.Lim, H.Yoo, B.H. Choi, J.H. Lee, H.-N. Lee, H.K. Lee, *Phys. Status Solidi* C 8, No. 3, 891 (2011).
39. B. H. Choi, Y. H. Lim, J. H. Lee, Y. B. Kim, H. N. Lee, H. K. Lee, *Microelectron. Eng.* 87, 1391 (2010).
40. Y.R.Shin, J.-H. Choi, W.-S. Kwack, S.-H. Kwon, 218th ECS Meeting, 2010.
41. A. Sherman, *Chemical Vapor Deposition*, Noyes Publications, New Jersey, 1987.
42. Y. J. Lee and S.W. Kang, *Electrochem.Solid-State Lett.* 5(10), C91 (2002).
43. M. Juppo, M. Ritala, and M. Leskela, *J. Vac. Sci. Techn. A15*, 2330 (1997).
44. P. Mårtensson and J-O. Carlsson, *J. Electrochem. Soc.*, 145, 2926 (1998).
45. R. Solanki and B. Pathangey, *Electrochem. and Solid-State Lett.* 3, 479 (2000).
46. A. Niskanen, A. Rahtu, T. Sajavaara, K. Arstila, M. Ritala and M. Leskelä, *J. Electrochem. Soc.* 152, G25 (2005).
47. L. Wu, E. Eisenbraun, *J. Electrochem. Soc.* 156, H734 (2009).
48. D-Y. Moon, D-S. Han, S-Y. Shin, J-W. Park, B. M. Kim, J. H. Kim, *Thin Solid Films* 519, 3636 (2011).
49. M. Juppo, M. Vehkamäki, M. Ritala, and M. Leskelä, *J. Vac. Sci. Techn. A16*, 2845 (1998).
50. J. W. Klaus, S. J. Ferro, and S. M. George, *Thin Solid Films* 360, 145 (2000).
51. S. M. George, J. W. Elam, R. K. Grubbs, and C. E. Nelson, *Mat. Res. Soc. Symp. Proc.* 672 (2001).
52. B. S. Lim, A. Rahtu, and R. G. Gordon, *Nature* 2, 749 (2003).

7

Organic and Hybrid Materials

7.1 Introduction

Inorganic materials have been the focus of the majority of the work on ALD. However, the same principles can be applied to the deposition of totally organic or partly organic materials. Although in conventional ALD many of the precursors contain organic ligands, during the ALD process these are removed to leave a completely inorganic film material. In contrast, in the deposition of organic materials, which is given the terminology molecular layer deposition (MLD), the resulting deposited layers retain these organic ligands or molecules in its structure. MLD shares the same characteristics as ALD: layer-by-layer growth, conformal coatings, ability to make nanolaminates, etc. We will divide our consideration of these materials into two sections, dealing separately with organic materials and with hybrid organic/inorganic materials.

7.2 Organic layers

The deposition of organic layers by MLD is relatively underdeveloped. However, it has the potential to fabricate a very large range of novel materials because of the vast range of organic molecules which are available for utilisation in the process. As well as the growth of the layers themselves, functionalization of the layer surface can be achieved by using the desired molecule in the final layer. There can be difficulties in obtaining bonding of the initial layers to a substrate but the initial attachment to the substrate can be controlled or enhanced by the application of a self-assembled monolayer to it.

Initial work was carried out on the deposition of polyamides such as nylon 66 [1] and polyimide [2]. However the material which has been studied ranges from these to polyazomethines [3], polyurethanes [4], polyureas [5] and terephthalates [6–7]. There are two regions of control in MLD deposition, (i) the molecular backbone itself and (ii) the chemical groups attached to the backbone which take part in the bonding between successive layers. The molecular backbone could be a single hydrocarbon chain, one containing aromatic carbon rings or a branched structure. The backbone could also contain various groups which will modify the polymer properties. In order to achieve good MLD growth, the end groups must be highly reactive but must form stable bonds when they react with each other. A schematic of the MLD process is shown in Figure 7.1

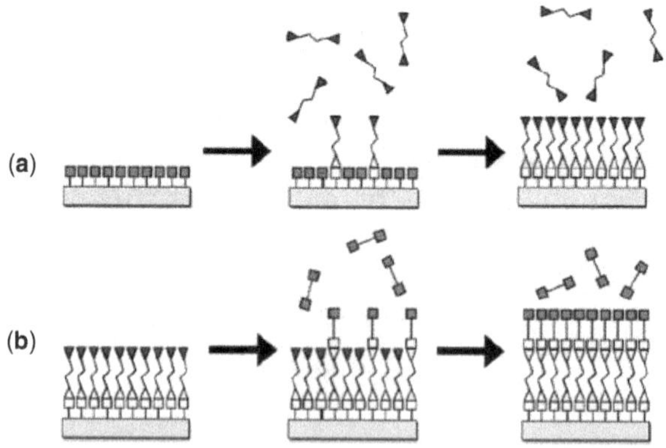

Figure 7. 1 Schematic of MLD growth using an AB binary sequence [8]. *Reprinted with permission from N. M. Adamczyk, A. A. Dameron and S. M. George, Langmuir 24, 2081, © 2008, American Chemical Society*

where the two precursors are applied in steps A and B leading to self-limiting growth.

As an example, the deposition of poly(*p*-phenylene terephthalamide) (PPTA) is described [6]. PPTA is a polyamide also known as Kevlar. The used precursors have been terephthaloyl chloride ($ClCOC_6H_4$-COCl) and the diamine is *p*-phenylenediamine ($NH_2C_6H_4NH_2$). The surface reactions for PPTA MLD have been proposed as following:

$$-NH_2^* + ClCOC_6H_4COCl \rightarrow -NHCO\ C_6H_4COCl^* + HCl \quad (7.1)$$

$$-NHCO(CH_2)_4COCl^* + H_2NC_6H_4NH_2 \rightarrow -NHCOC_6H_4CO{-}NH(CH_2)_6NH_2^* + HCl \quad (7.2)$$

where the asterisk designates the surface species. These reactions of PPTA MLD are also shown in the Figure 7.2.

Patterning of a polymer MLD layer can be achieved by selective deposition by changing the hydrophilicity/hydrophobicity of the substrate in selective areas. This has been done using a combination of hexamethyldisilazane (HMDS) exposure for the hydrophilic regions and dichloromethane for the hydrophobic regions [9].

7.3 Hybrid Organic-inorganic Layers.

As well as wholly organic materials, it is possible to create ALD processes combining a metal precursor and an organic molecule to produce alternating metal-organic or hybrid film structures. Many of these are based on the normal precursors used for metal oxide such as trimethyl aluminum. The combination of this with a diol material such as ethylene glycol results in what is known as "alucone" by analogy with silicone [10]. By a similar process, other compounds can be produced such as zincones [11], titanicones [12] and zircones [12] known collectively as "metallocones".

As an example, the reaction of TMA with ethylene glycol proceeds as follows:

$$AlR^* + HOR'OH \rightarrow AlOR'OH^* + RH \quad (7.3)$$

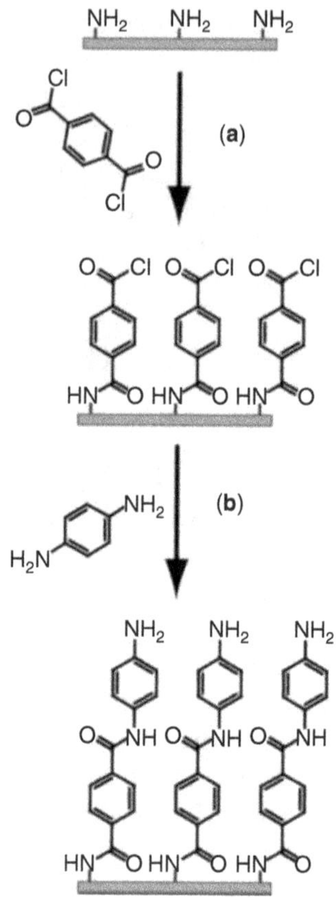

Figure 7.2 Illustration of surface chemistry for PPTA MLD. In the step A, terephthaloyl chloride reacts with amine surface groups to form an amide linkage. In the step B, the diamine reacts with acid chloride surface groups to form an amide linkage [6]. *Reprinted with permission from N. M. Adamczyk, A. A. Dameron and S. M. George, Langmuir 24, 2081 © 2008, American Chemical Society.*

$$AlOR'OH^* + AlRx \rightarrow AlOR'OAlR_{x-1} + RH \qquad (7.4)$$

R in this case is the methyl group and R′ is C_2H_4.
The reactions can be pictured as shown in Figure 7.3.

The growth of the metallocones has the normal ALD characteristics – thickness proportional to the number of cycles and growth per cycle independent of the supply of precursor (above the saturation level). However, in this case there is no "ALD temperature window" where the deposition rate is stable over a range of temperature. In contrast, typically the deposition rate decreases with

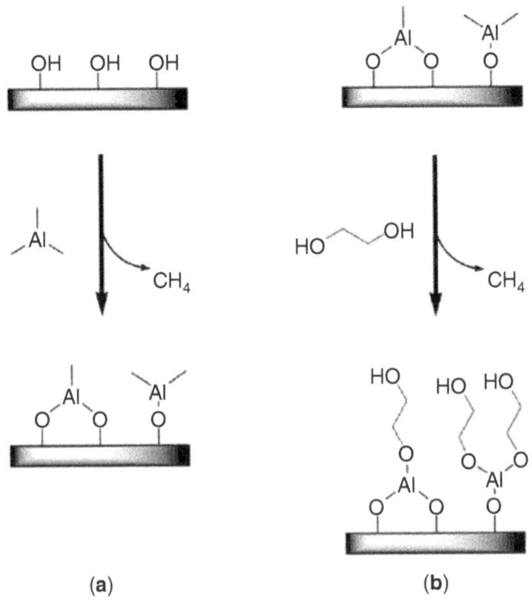

Figure 7.3 Surface reactions during the formation of an alucone using TMA and ethylene glycol as precursors [13]. *Reprinted with permission from S.M. George, B. Yoon and A.A. Dameron, Acc. Chem. Res. 42, 498 © 2009, American Chemical Society.*

reaction temperature [13]. This behaviour also occurs in purely organic layers [6]. This existence of the same reactive group at both ends of the organic molecule (e.g. ethylene glycol HOC_2H_4OH) can lead to the situation where both ends react with the active surface groups so terminating the growth at certain points and probably leading to porosity in the films. This can perhaps be overcome by using molecules with different end groups so that only one will react during a particular step of the process. one postulated process. One such process is the "ring opening" reaction. One example of this reaction uses a cyclic azasilane which can react with hydroxyl groups on the substrate surface. The silicon in the azasilane form a bond with the oxygen in the OH group and the ring structure opens giving a linear molecule terminated in an amine bond which will not react with any OH groups on the surface. If the second precursor is also a ring structure such as a cyclic carbonate, this will react with the amine, again opening the ring and exposing another OH bond at the end of the polymer chain. Figure 7.4 shows this reaction process.

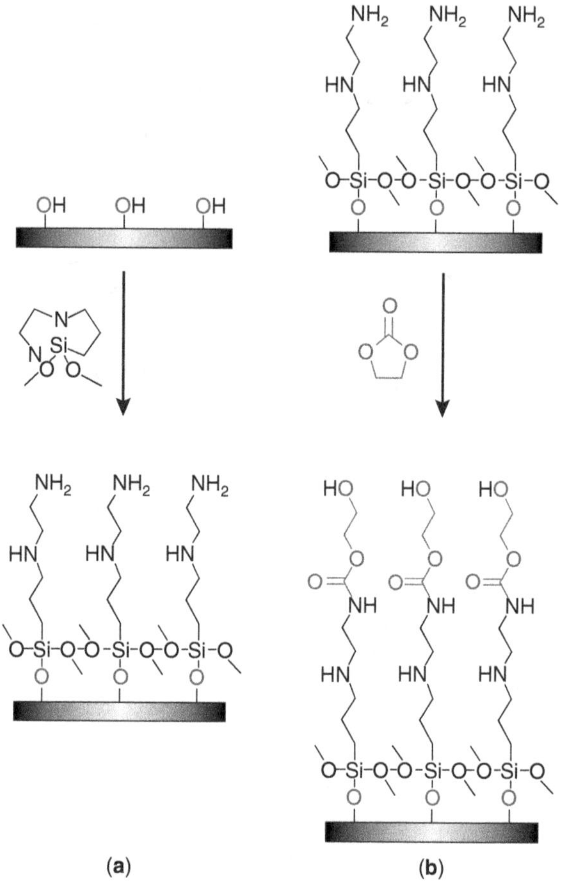

Figure 7.4. Ring opening reaction steps using an azasilane (step A) and cyclic carbonate (step B) [13]. *Reprinted with permission from S.M. George, B. Yoon and A.A. Dameron, Acc.*

There are other possible schemes for depositing metallocones, including a three step ABC process. The interested reader is referred to the review of George [13], for example, for a detailed account.

7.4 Applications of Organic and Hybrid Films

The applications of MLD organic and hybrid films are still in their infancy. The proposed applications are where the characteristics of ALD/MLD are important such as conformality, uniformity, thickness control and lack of defects. With inorganic ALD films, these features are being exploited to produce high quality insulators, diffusion barriers and corrosion protection layers, for example. However, on flexible substrate materials, the brittle nature of the ceramic ALD films makes them prone to damage on bending. One aim is to interpose polymeric layers between the ceramic layers or to utilise the hybrid nature of the metallocones to increase the strain at which cracking occurs or to increase the diffusion path length in gas barrier layers [14]. For these purposes, the metallocones should have flexibility. Preliminary results of measurements of the mechanical properties indicate that alucones have a greater elastic modulus than conventional polymers [14], are more brittle [14–15] and are liable to change their properties on exposure to the environment [10–14]. However, the enormous range of organic materials which can be used in MLD, both for hybrid and for organic layers, gives plenty of opportunity for improvements in these properties and for development of real applications.

References

1. H. Shao, S. Umemoto, T. Kikutani and N. Okui, *Polymer* 38, 459 (1997).
2. T. Yoshimra, S. Tatsuura and W Sotoyama, *Appl. Phys. Letts.* 59, 482 (1991).
3. T. Yoshimura and Y. Kudo, *Appl. Phys. Express* 2, 015502 (2009).
4. J.S.Lee, Y.J. Lee, E.L.Tae, Y.S. Park and K.B.Yoon, *Science* 301, 818 (2003).
5. A. Kim, M.A. Filler, S. Kim and S.F. Bent, *J. Am. Chem. Soc.* 127, 6123 (2005).
6. N. M. Adamczyk, A. A. Dameron and S. M. George, *Langmuir* 24, 2081 (2008).
7. T. V. Ivanova, P. S. Maydannik, D. C. Cameron, *J. Vac. Sci. Technol. A* 30, 01A121 (2012).
8. Y. Du and S. M. George, *J. Phys. Chem. C*, 111, 8509 (2007).
9. T. Yoshimura,N. Terasawa, H. Kazama, Y. Naito, Y. Suzuki and K. Asama, *Thin Solid Films* 497, 182 (2006).
10. A.A Dameron, D. Saghete, B.B. Burton, S.D. Davidson, A.S. Cavanagh, J. A. Bertand and S.M. George, *Chem. Mater*. 20, 3315 (2008).
11. Q. Peng, B. Gong, R.M. VanGundy and G.N Parsons, *Chem. Mater*. 21, 820 (2009).
12. S.M. George, B.H. Lee, B. Yoon, A.I. Abdulgatov and R.A. Hall, *J. Nanosci. Nanotechnol.* 11, 7948 (2011).
13. S.M. George, B. Yoon and A.A. Dameron, *Acc. Chem.* Res. 42, 498 (2009).
14. D. C. Miller, R. R. Foster, S-H. Jen, J. A. Bertrand, D. Seghete, B. Yoon, Y-C. Lee, S. M. George and M. L. Dunn, *Acta Materialia* 57, 5083 (2009).
15. D. C. Miller, R.R. Foster, Y.D. Zhang, S-H. Jen, J.A. Bertrand, Z.X. Lu, et al., *J Appl. Phys.* B105, 093527 (2009).

8

ALD Applications and Industry

8.1 Introduction

In recent years two new technology developments have occurred that are now converging. In the field of microelectronics, the pressure to reduce the cost of manufacture of integrated circuits has become extremely important as design rules require deposition of ever thinner films (dielectric, conductor, and semiconductor materials). Since these films (nanometer thick) must still be coated on large silicon wafers (300 mm or 3×10 [8] nanometers), the challenge has been to discover ways to deposit these films so that they are extremely uniform. Because of the enormous financial implications of these developments considerable development of ALD has occurred, as this technology is one of the ways these advanced designs can be realized. One example is Intel's 45 nm CMOS process technology launched in 2007, where hafnium based gate oxide is made by ALD [1].

The other technology, referred to as Nanotechnology, relates to the recognition that many devices (not just microelectronic devices) can be manufactured at extremely low cost when dimensions are reduced to the nanometer scale. Probably the first area that received attention was the development of Microelectromechanical and nanoelectromechanical systems (MEMS/NEMS). As an illustration, an accelerometer can be fabricated from a simple cantilever. If fabricated on the nanometer scale, the cost of such a device can be reduced dramatically. Since every air bag in every car manufactured worldwide uses an accelerometer to sense a crash, we can see the huge market implications of such developments. To foster development of such Nanotechnology applications, commercial and government organizations are providing extensive development funding.

Historically, CVD has been the method of choice for deposition of two of the most important thin films (SiO_2 and polysilicon) used in semiconductor fabrication. This is because in the early years the required film thicknesses were fairly large (1–2 microns), and high deposition rates were critical. Today there are applications for ALD films, but their deposition rate is quite low compared to those for CVD processes. In spite of this, ALD is being actively pursued, because of its unique properties when deposition rate is less important, as for nanometer thick films.

One feature of Nanotechnology devices that presents some challenge to other ways of depositing thin films is the need, in some applications, for nanolaminar multi-layer films. As we have seen in the earlier chapters such nanolaminates can be readily grown with the ALD process. Another very important feature of many Nanotechnology devices is their topography. In other words, whether it be a MEMS/NEMS device, a highly porous catalyst with a high surface area relative to its volume, or other high aspect ratio system, it is very important to have a deposition technique that creates extremely conformal films. Again this feature can be readily achieved by the ALD process, and is very hard to accomplish by other deposition processes.

Since the manufacturing processes needed to manufacture new Nanotechnology devices are already in use in the semiconductor industry, the eventual use of ALD processes for such applications is very likely. Especially since this new process technology can deposit films not only with extraordinary conformality and nanometer thickness control, but at lower temperatures than conventional microelectronic processing. Actually, a substantial percentage of the equipment being sold for use in Nanotechnology manufacturing comes from the used equipment market.

ALD has been exploited in industrial applications since the early 1980s in the manufacturing of thin film electroluminescent displays. Thereafter ALD has found new industrial applications outside of more conventional areas, where protection of silver jewelry is a good example. A common feature in early-developed and new ALD applications is the capability of controlling the film thickness uniformity and conformality on the nanometer scale.

In the present chapter we will review several Nanotechnology applications that are benefiting from or have great potential for the use of ALD processes. Since the number of such potential applications is extremely large, we will only be able to touch on the most important ones briefly.

8.2 MEMS/NEMS

For the purposes of this review, we will define a MEMS/NEMS device as any nanometer scale machine where there is an interaction between a mechanical motion and an electrical signal. By mechanical motion we will include not only motion of solid materials but also flow of fluids. In one application, an accelerometer will detect acceleration or deceleration by sensing the motion of a mechanical part, with the aid of an electrical signal. For example, nanometer scale motion can be sensed electrically by the closing of a reed switch, or as a change in capacitance between two mechanical parts. In another application, an inkjet head can be built where local heating of a liquid can vaporize it and cause trapped liquid to be expelled through an orifice. Working within this definition, we will review a number of commercially important applications.

To begin with we can review the typical procedure for fabrication of such devices. Initially, these devices were built on silicon wafers to take advantage of well-known fabrication techniques and existing reactors. Silicon dioxide can either be grown thermally, or deposited by CVD if thick films (> 1 μm) are called for, and any required etch patterning carried out. Next, a thin film of conformal polysilicon is deposited by CVD. Coating this polysilicon with CVD silicon dioxide, and again patterning the oxide by etching creates the desired structure. It is important to note that CVD polysilicon process is unique in that the coating is very conformal. Also, the silicon dioxide is readily etched either with hydrofluoric acid (a wet isotropic etch), or with reactive ion etching (a dry and anisotropic etch). Since the wet etch is isotropic, areas under the polysilicon parts may be etched. This is referred to as releasing the polysilicon part. The polysilicon is typically patterned with a reactive ion etch.

There are many references detailing the fabrication processes used, so we will make no attempt to go into this aspect of MEMS/NEMS device manufacturing techniques. However, the degree of complexity achievable is illustrated in Figure 8.1 [2]. As can be seen, the active parts (gears) are free standing, and this has been enabled by wet etching the silicon dioxide layer used to pattern the device. In order to realize this level of complexity, five levels of polysilicon and oxide were employed.

The rationale for the anticipated future application of ALD to MEMS/NEMS devices lies in the fact that only the CVD polysilicon has the ability to deposit conformally. Deposition of any other films (metals, nitrides, oxides, etc) by CVD will generally be non-conformal to a considerable degree. ALD deposition offers conformal films of almost any material, so that MEMS/NEMS designers can have additional flexibility in the design of advanced devices.

As noted earlier, one of the first applications of nanotechnology was for the development of MEMS accelerometers. Initially, accelerometers used for car airbags were built from macro parts, cost about $15 each and were bulky as well. For some years now, accelerometers have been fabricated by techniques derived from integrated

Figure 8.1 A MEMS ratcheting system showing a 100 μm diameter gear [2]. *Courtesy of Sandia National Laboratories, SUMMiT(TM) Technologies, www.mems.sandia.gov*

circuit manufacture, and cost on the order of \$1–\$2. An example of a car airbag accelerometer [3] that operates with many interdigitated fingers and uses electrostatic sensing is shown in Figure 8.2.

As these MEMS structures have become more complex additional sophisticated applications have been developed successfully. One of these is the MEMS gyroscope [4] used for angular rate sensing shown in Figure 8.3. This device built from polysilicon has an 8 μgm mass suspended 2 μm above a substrate. It has 5000 fingers spaced 1.6 μm apart, and it is so sensitive that it can detect deflections as small as 0.00016 Å.

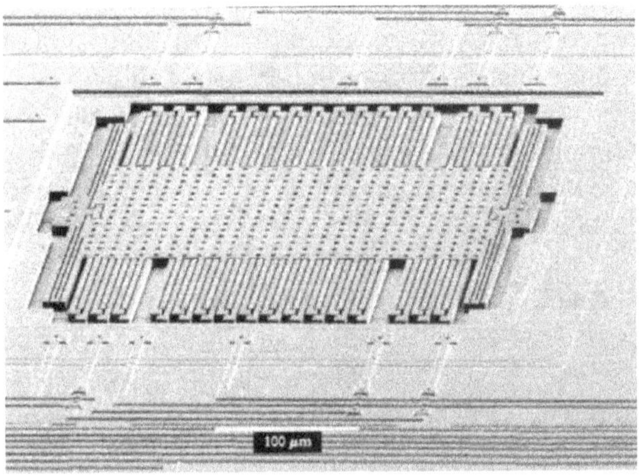

Figure 8.2 A MEMS accelerometer for use in car airbag applications [3]. *Copyright Analog Devices, Inc. All rights reserved.*

Figure 8.3 MEMS gyroscope for angular rate sensing applications [4]. *Copyright Analog Devices, Inc. All rights reserved.*

The MEMS gyroscope is now being used in automobiles for anti-rollover control, GPS navigation, and electronic stability control. They also are used in a number of defense and aeronautical applications. Finally, they are seeing use in consumer applications, such as camera stabilization, and video games.

Another large scale application of MEMS/NEMS technology occurs in pressure sensing. It has become possible to reduce the cost and size of such sensors so that they are now being installed in the tires of all new cars.

In fact, it is safe to assume that any macro fluid device (i.e. flow valves, pressure regulators, mass flow controllers, pumps, etc.) can and will be developed in the MEMS/NEMS world. This technology is often referred to as microfluidics, but really is a subset of MEMS. The number of potential applications in the microfluidics area is expanding very rapidly [5], and includes the life sciences, the pharmaceutical industry, and agriculture. The most fully developed application, however, is in the manufacture of ink jet heads for printers.

One approach, pioneered by Hewlett-Packard, has been to create a reservoir of ink within a cavity formed in silicon. The cavity then has a small hole that allows an ink droplet to escape. On a portion of the inner surface of this cavity a small thin film resistive heater is formed. When an electric current is applied, the resistive heater vaporizes some of the fluid ink and this causes a drop of liquid ink to be ejected. This process is repeated 12,000 times per second. As should be anticipated, the thermal efficiency of such a design will be small ($\ll 1\%$), so other approaches have been developed, and are in commercial production. For example, if a piezoelectric material is placed adjacent to the ink cavity it can be electrically activated so that the cavity shape can be modified very quickly. This change of shape will also eject an ink drop.

The heater used in the ink jet heads can be as thin as 200 Å, and the reason that it is difficult to form is that it has to be physically and electrically insulated from the substrate. In addition, it has to be protected from corrosion by the ink, and as well from cavitation damage caused by collapsing ink bubbles.

Details of the thermal ink jet head fabrication are described in two patents issued to HP [6]. Thermal ALD is mentioned as one of the possible techniques for fabrication of such devices. In addition, Samsung has applied for a US patent [7] for another method of fabricating a thermal ink jet head resistive heater incorporating several noble metals. In this case, Radical Enhanced ALD is referred to as one of the ways to fabricate such a structure.

Another major commercial application of MEMS technology is in the fabrication of optical MEMS (MOEMS). For example, an optical scanner can be built using a very small mirror that can be moved by some electrical pulse. This mirror would then reflect a light beam from a laser diode onto a bar pattern. By moving the mirror the light beam can be made to traverse the bar pattern, and the reflected light will be converted into a digital readout of the bar pattern. In such a device the actuation is effected by using an electric current through a very small heater to distort a cantilever beam holding the mirror.

The actuator is fabricated from nanometer thick layers using materials traditionally used in integrated circuit fabrication. Also, it has a three dimensional shape [8]. Accordingly, if at some future date a device similar to this one requires the use of different materials, and/or a conformal coating, then ALD processes may be pressed into service.

Far and away, the largest commercial product using MOEMS is the Digital Light Processing (DLP) technology developed by Texas Instruments about 20 years ago [9]. This technique uses an array of aluminum micromirrors or Digital Micromirror Devices (DMD) that are manufactured in a traditional integrated circuit fab, and that are used to produce a very high performance digital projector. Each DMD pixel is 16 μm square, and there are over a million on each DLP chip. Each can be rotated independently by electrostatic forces created when a voltage is imposed by a CMOS device built into the base of the chip.

In Figure 8.4 two assembled DMD pixels are shown, where the aluminum mirrors are portrayed as transparent in order to show the mechanism clearly [9]. An exploded view of a single DMD pixel is shown in Figure 8.5.

As with all MEMS devices, the chip topography is very complex, and it is fabricated using a number of different materials that all have to be deposited correctly.

Another device, built with MOEMS technology is a reflective optical display developed by Quallcom [10], that uses a deformable membrane to adjust the optical depth of a cavity so that a primary light wave (red, green and blue) can be either absorbed or reflected. The membranes movement is the result of an electrostatic force.

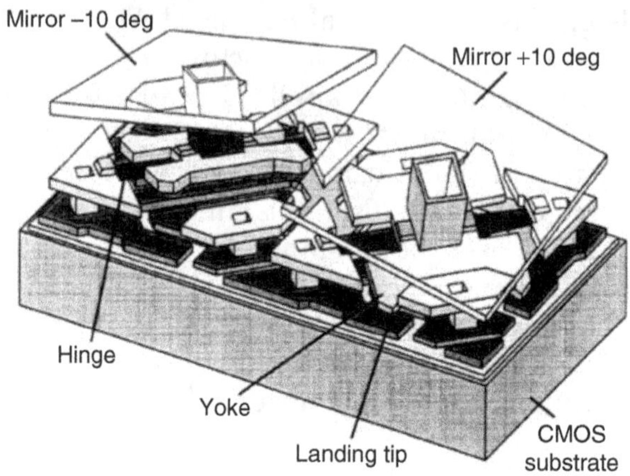

Figure 8.4 Two Digital Micromirror pixels [9]. *Reprinted with permission from SPIE, © 1996.*

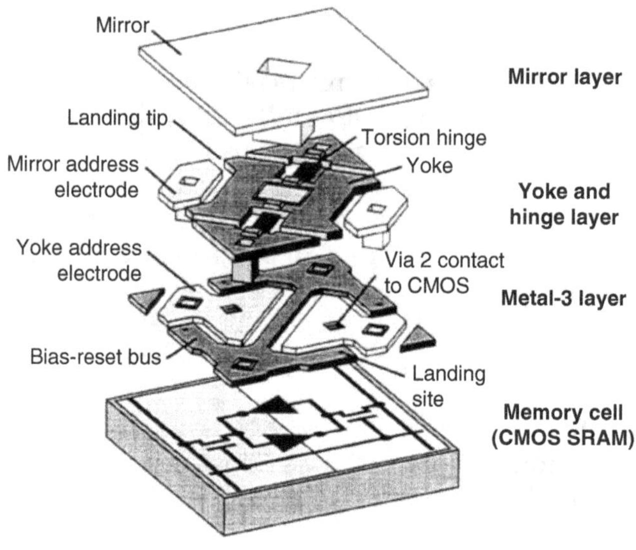

Figure 8.5 Exploded view of a pixel [9]. *Reprinted with permission from SPIE, © 1996.*

There are many other interesting MOEMS technologies, but they are far too numerous to be covered in this brief review. Before leaving the topic of MEMS/NEMS, we should mention several other MEMS/NEMS applications where ALD can play a role. For example, passive electrical devices such as switches, capacitors, inductors, can also be miniaturized. When such parts are combined on a chip RF circuits can be built. These devices are typically referred to as RF MEMS and RF NEMS. In one example, the fabrication of a MEMS capacitive switch was explored where it was fabricated by thermal ALD or Radical enhanced ALD [11].

Another MEMS product being developed is the microminiaturized microphone [12]. This product is manufactured in silicon in a standard CMOS fab environment, ensuring efficient high volume manufacturing. It consists of an air cavity etched into silicon covered by a membrane that vibrates as the air pressure changes in the cavity. Because the microphone can be fabricated in silicon with a standard CMOS process, an entire chip with all the required electronics to control it can be integrated into a single silicon chip.

Finally, we should note that when a MEMS/NEMS device involves either rotary or linear sliding contacts, then friction and wear are significant concerns. One way to mitigate such effects would be to coat all surfaces with a material that reduces friction and/or limits wear. Since these structures may have surfaces that need coating that are on the underside of the parts, a coating process that gives extremely good conformality should be very valuable. As we have noted many times, ALD processes can have close to 100% conformality, therefore deposition of films such as TiN (to reduce friction) and Al_2O_3 (to reduce wear [12–13]) are clearly feasible.

8.3 Thin Film Magnetic Heads

Thin film magnetic heads (TFMH's) to read and write data into hard disc drives have been one commercial application of ALD for more than ten years. Due to the development in microelectronic devices and need for downscaling of magnetic heads, the conventionally used sputtering technology has reached the limits of production of high quality insulator layer on increasingly three dimensional structures [14]. In this application ultrathin ALD Al_2O_3 grown from TMA and H_2O was able to meet the requirements that were needed for the development of magnetic head. Other possible materials reported are AlN [15] and nanolaminates of Al_2O_3 and SiO_2 [16].

8.4 Coating Nanoparticles, Nanomaterials and Porous Objects

Coating on nano- and micron sized particles, nanomaterials and other fine objects is an interesting field with great commercial potential for ALD. Conformal and uniform nanometer scale coating will provide opportunity to bring a variety of functionalities to these objects. Conventional thin film and coating methods, such as CVD and sol-gel, used in particle coating suffer for example from particle agglomeration and non-uniformity of the coating. However fluidized bed ALD has a proven capability to deposit conformal and uniform coatings on a number of different nanoparticles and fine objects [17] , and this has also opened up new commercial applications to ALD [18].

Fluidized bed reactors may have some difficulties related to the particle properties and consequently the ALD chemistry. When particle size decreases below 1μm the overall surface area of particles increases radically. This requires larger quantities of reactants to get a complete reaction over the particle surface. The fluidized bed reactor may suffer low efficiency if the reactant has a low reactive sticking coefficient and the residence time in the reactor is not sufficient for the reactant to react with the whole particle surface area. Low efficiency can be improved by allowing longer lifetimes of reactant in contact with the particle surface area. Longer lifetime can be achieved with static reactant exposure, which, however, cannot be obtained in a fluidized bed rector where the fluidization is based on the presence of reactant gas and additional gas flows. However, static reactant exposures simultaneously with fluidization of particles can be arranged in a rotary ALD reactor, which is another approach for particle ALD [19].

Particle ALD has been studied for many different applications [17]. To give a few examples here, one recent study showed that electrochemical performance of $LiMn_2O_4$ cathode material for lithium ion batteries improved by using ultrathin ALD ZnO (1.02 nm) to provide a shield to manganese ion dissolution and also isolating $LiMn_2O_4$ from the electrolyte [20]. An earlier study demonstrated the ability to combine the properties of organic and inorganic material by depositing a layer of Al_2O_3 on HDPE polymer particles prior to extrusion to form nanocomposite membranes [21].

Feasible applications for particle ALD can also been found in prevention of environmental pollution. TiO_2 is widely used in industry as a photocatalytic material because it is inexpensive, safe to use

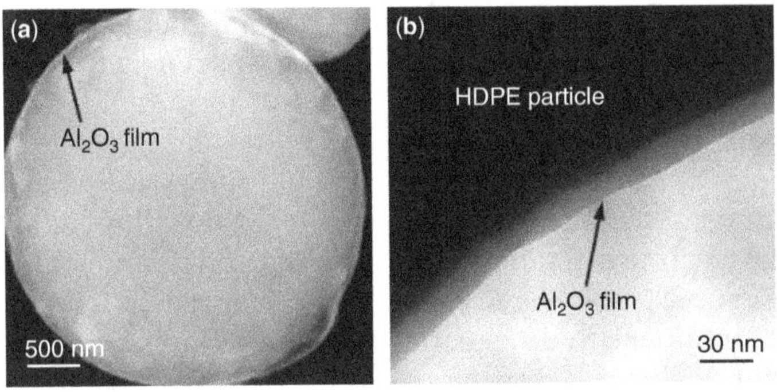

Figure 8.6 ALD Al_2O_3 coating on HDPE particle [21]. *Reprinted from J. Membrane Sci. 322, X. Liang, D. M. King, M. D. Groner, J. H. Blackson, J. D. Harris, S. M. George, A. W. Weimer, Barrier properties of polymer/alumina nanocomposite membranes fabricated by atomic layer deposition,105 © 2008 with permission from Elsevier.*

and effective in different environments. Further improvement of photocatalytic activity of TiO_2 is desired to increase its efficiency and wider usage. One recent study has shown that it can be done by depositing Pt clusters on TiO_2 particles by ALD [22]. This may have high impact in heterogeneous catalysis processes whether they operate in particle or in porous material media.

There are many applications that make use of porous materials with nanometer scale pores. In many cases, the functionality of such materials can be improved by coating them with nanometer thick films. One technique for depositing such coatings on porous structures is ALD. In this section we will review several applications that may prove to be significantly enhanced by the use of this new processing technology.

Catalysis has been in use for many years, in a number of industrial chemical manufacturing processes such as synthesis of ammonia NH_3 from H_2 and N_2. Also, oxidation of CO by O_2 in order to prevent CO emissions from automobile exhausts is an important application [23]. These processes are referred to as heterogeneous catalysis where a solid catalyst promotes a gas phase reaction, but remains unchanged. A number of materials can function as catalysts, such as iron for ammonia synthesis and platinum for CO oxidation. Their cost, in particular for the noble metals (Pt, Ru, Pd, etc.), that turn out to be among the most effective catalysts, can be very high. However, all that is needed for a catalyst to function is for there to be a thin layer of material on a support surface. Since the catalyst only serves to promote chemisorption of the reactants being catalyzed, and is not consumed in this reaction, several monolayers may be sufficient. This should allow significant cost savings to be realized when an atomic layer deposition process is used rather than some other method of thin film deposition.

A number of studies have examined the successful ALD deposition of several metals, such as Ti and Ni, onto different supporting sintered porous ceramics, such as silica and alumina [24–25]. In addition, some results have been reported for the use of nickel as a catalyst for toluene hydrogenation [25]. Since this is a commercially important application, it remains an active field of research [26–27].

Another important application for deposition of conformal nanometer thick layers, including catalytic materials, on porous structures is in fuel cells that use solid electrolytes. A great deal of research has been done over the last 5–6 years on the development of polymer electrolyte membrane and direct methanol fuel cells (DMFC) [28]. Similar research has been reported for many years on the development of solid oxide fuel cells (SOFC) [29]. ALD research has been active in this field for several years [30,] [31–34]. ALD has enabled the significant reduction of electrode material in Pt anodes [32] and also the thickness of Yttria Stabilized Zirconia electrolytes used in SOFC [34].

Finally, porous inorganic membranes that are used to separate gases will clearly be important for hydrogen purification, as we proceed towards a hydrogen-based economy [35]. In addition, such membranes that have to operate under extreme conditions such as high temperatures or corrosive gases will also be very valuable. Under the circumstances where the best porous membranes available have relatively large average pore sizes (> 50Å), one can use ALD to grow conformal nanometer thick films of the same material and reduce the average pore size. In one recent study [36], it was demonstrated that SiO_2 or Al_2O_3 could be grown on porous alumina to the extent that the average pore size could be reduced to molecular diameters.

8.5 Optical Coatings

There are many instances in which thin films are used to coat optical components, such as mirrors, or to fabricate thin film displays. There are obviously a number of ways to deposit these films, such as evaporation, sputtering, CVD, etc. However, frequently the parts being coated require very carefully specified characteristics, and these deposition techniques all have their shortcomings. In particular, sputtering and evaporation require expensive deposition systems, and even then it is difficult to grow films with precise thickness uniformly over substantial areas. There ALD can offer its unique characteristics.

In one recent study, thin films of TiO_2 and Al_2O_3 were studied for use as coatings intended for an inertial fusion experiment to generate fusion energy with high-power lasers [37]. Those parts that would need coatings would be large mirrors, polarizers and band-pass filters. ALD deposition techniques we have already reviewed in Chapter 4 were used with $TiCl_4$, TMA and water. Depositions were carried out from 25°C to 400°C.

A two-layer antireflection coating was fabricated at 200°C with a first TiO_2 layer 196 Å thick, and a second Al_2O_3 layer 1,111 Å thick, intended to transmit at 532 nm. When this film was applied to both sides of a silica substrate the transmission was 96.5%. Also, film thickness variation was less than 1 % over a 25 cm substrate. Based on these results, the authors conclude that this film would be usable for polarizers and band-pass filters.

Additional experiments showed that either Al_2O_3 or TiO_2 grown at 25°C were amorphous, while either grown at higher temperatures was crystalline. Interestingly, the amorphous film had a much higher resistance to laser damage than the crystalline films. Although these low temperature films had some chlorine contamination that did not seem to affect their ability to resist laser damage. Therefore, the amorphous form of these films would likely be the choice for this application.

A second study used ALD to deposit ZnS:Mn, Al_2O_3, and ZnO:Al in the fabrication of an electroluminescent display [38]. The particular structure used is shown in Figure 8.7. The device is built on an alumina substrate with a Pt/Ag contact layer and a $BaTiO_3$ dielectric layer both fabricated by screen printing. The $BaTiO_3$ layer was thick, 46 μm, and this screen printed layer was quite rough. In fact, this roughness was on the order of the subsequent thin film thickness (~ 1.12 μm). For this reason, it was found that line of sight deposition techniques, such as evaporation or sputtering gave poor results when compared to the highly conformal ALD coatings. Because of the ability to incorporate the thick $BaTiO_3$ layer, an increase in brightness of over 2:1 compared to a traditional electroluminescent display design could be

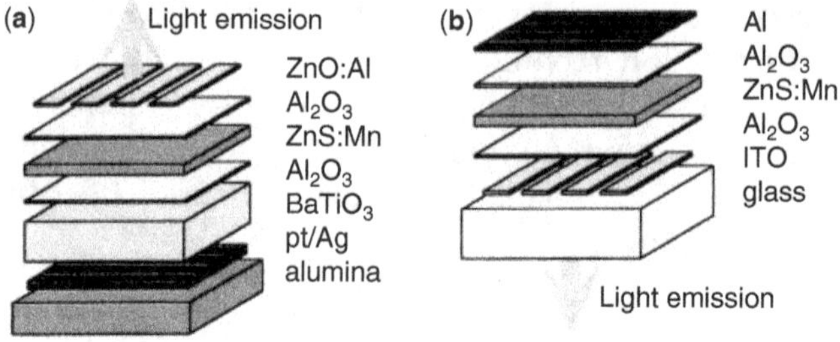

Figure 8.7 Hybrid electroluminescent device [38]. © *2002 The Japan Society of Applied Physics.*

achieved. Also, the ALD process enabled a substantial improvement in uniformity of light emission over that achievable with evaporated layers.

Optical coatings by ALD has also enabled commercialization of infrared cut-filters [39] and UV polarizers [40].

8.6 Thin Film Electroluminescent Displays

Electroluminescent devices are an ideal example of an ALD application since they were the initial motivator for ALD development. Tuomo Suntola and co-workers were granted a patent in 1977 and ALD was further developed for industrial manufacturing of thin film electroluminescent (TFEL) displays [41–42] A TFEL display is comprised of a layered structure of electrodes composed of ALD grown luminescent manganese-doped zinc sulphide and oxides as shown in Figure 8.8 (left) [43].

Initially the phosphorous ZnS:Mn layer for monochrome and multicolor (red-green-yellow) displays were grown by using $ZnCl_2$, $MnCl_2$ and H_2S as the reactants at 500°C. Later, lower deposition temperature below 400°C was allowed with higher vapor pressure reactants like diethyl zinc (DEZ) and $Mn(thd)_3$. Development of color EL displays has widened the material combinations in the phosphorous layers. ZnS:Mn/SrS:Ce, ZnS:Tb, ZnS:Sm and CaS:Eu ALD processes have been developed. In TFEL devices the insulation of luminescent layer has a key role. A widely used insulator is a mixed Al_2O_3/TiO_2 structure. Other oxide materials, such as Ta_2O_5, ZrO_2, HfO_2 and their mixtures for insulators in the TFEL structure have been studied [43].

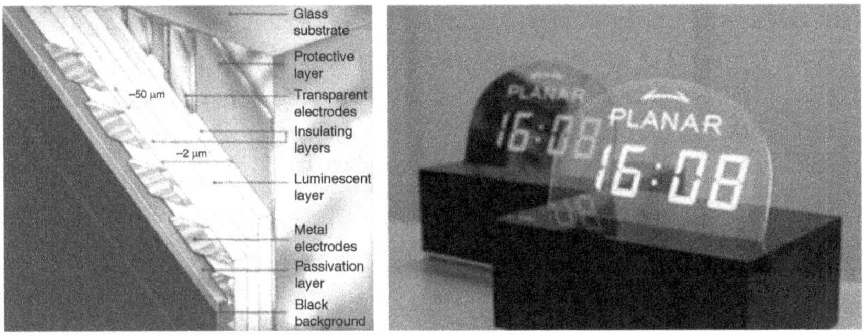

Figure 8.8 Schematic representation of thin film electroluminescent display structure (left) and image of transparent electroluminescent display (right). *Copyright Planar Systems Inc. 2012.*

8.7 Solar Cells

The direct conversion of solar energy to electricity has been an active field of research for many years now. In recent years it has achieved some commercial success, but further performance advances are needed for it to achieve its full potential [44]. Many of the recent improvements in this technology depend on the creation of nanometer thick thin films with unique qualities. Fabrication of such films by ALD will likely be just one additional way to create them, in addition to techniques such as sputtering, CVD, etc.

Examples of the application of ALD to improve the performance of solar cells involves the use of thin ZnO layers formed from DEZn and water [45–46]. Also see Chapter 4. In one system, an amorphous silicon solar cell is coated with boron-doped ZnO which is used as a transparent conducting oxide [45]. The critical feature of the ALD process is its ability to conformally coat an initial 1 µm textured high resistivity ZnO(B) layer (deposited by CVD) with a lower resistivity 0.4 µm ZnO(B) layer. In this way the transparent conducting layer was able to retain its texture as well as realizing a lower resistivity.

The second application involved the use of ALD ZnO, Zn(O,S) or (Zn,Mg)O as a high resistance buffer in a Cu(InGa)Se_2 (CIGS) solar cell [46–50]. The usual structure of such solar cells is ZnO/CdS/CIGS/Mo/glass. However, the thin high resistivity CdS buffer layer is deposited by a wet process and the hazardous nature of Cd have opened the window for ALD to overcome these issues. ALD is compatible with other vacuum process steps in fabricating the CIGS cells and is capable of depositing multicomponent thin films with accurate thickness and structure. CIGS devices with pure or oxygen rich ZnO suffer from low open circuit voltage, whereas devices with a ZnS buffer layer from very low photocurrent. A mixed Zn(O,S) buffer layer is therefore needed [48]. For example a mixed ALD Zn(O,S) layer with a 5:1 O:S ratio is the result of 5 cycles of H_2O pulse/N_2 purge/DEZ pulse/N_2 purge for every H_2S pulse/ N_2 purge/DEZ pulse/N_2 purge cycle. This ratio was also found to result in the best cell efficiency which was comparable with the efficiency of CdS reference cell [50]. An alternative buffer layer for Zn(O,S) is (Zn,Mg)O. In one study the film was grown following an ALD cycle similar to the above Zn(O,S) but for MgO deposition the ALD cycle was $MgCp_2$ pulse/N_2 purge/H_2O pulse/ N_2 purge. The desired Zn/Mg ratio can be obtained by varying the number of cycles of ZnO and MgO steps. Figure 8.9 shows the CIGS cell characteristics as a function of ZnO/MgO ratio [49].

An ALD buffer layer for CIGS solar cells has also gained industrial interest, see Section 8.9. Beneq Oy (Finland) offers the commercial industrial size ALD tool TFS 1200, available for 1200×1600 mm^2 panels based on Zn(S,O) or (Zn,Mg)O processes. A sample glass coated with Zn(S,O) ALD layer is shown in Figure 8.10 [51].

Surface passivation of crystalline silicon (c-Si) solar cells by ALD has been increasingly studied for past few years [52–56]. This exists as an application and availability in production systems in ALD

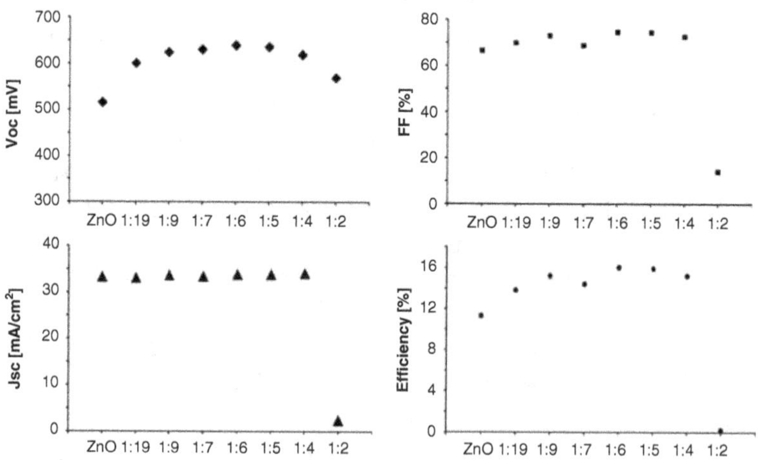

Figure 8.9 Figure 8.9. $CIGS_2$ solar cell parameters as a function of Mg content in (Zn,Mg)O buffer layer deposited by ALD [49]. *Reprinted from Thin Solid Films 515,C. Platzer-Björkman, T. Törndahl, A. Hultqvist, J. Kessler, M. Edoff, Optimization of ALD-(Zn,Mg)O buffer layers and (Zn Mg)O/Cu(In,Ga)Se_2 interfaces for thin film solar cells, 6024 © 2007 with permission from Elsevier.*

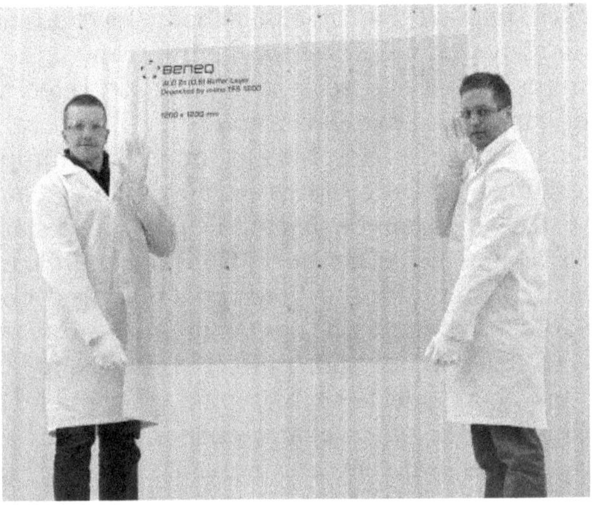

Figure 8.10 ALD Zn(O,S) buffer layer on 1200×1200 mm^2 substrate manufactured by Beneq [51]. *Photo courtesy and Copyright Beneq Oy 2012.*

industry as well, see Section 8.10. It has been demonstrated that cell efficiency can be increased by an Al_2O_3 layer that has a high density of negative built-in charges. With the use of Al_2O_3, the minority carrier concentration in the silicon can be reduced which then reduces the recombination rate. Surface passivation by ALD Al_2O_3 layer also allows wider base material selection since both highly doped p-type and n-type c-Si can be effectively passivated [55]. Industrial interest towards c-Si passivation can be seen through ALD tool and process development enabling systems with high wafer throughput and consequently low deposition cost per wafer [57–62].

Finally, there is interest in developing Cu_xS solar cells with nanoporous TiO_2 electrodes [63]. The primary difficulty in fabricating such cells is the conformal deposition of the Cu_2S film within the pores of the TiO_2 layer. In this study ALD from $Cu(thd)_2$ and H_2S at 175°C yielded $Cu_{1.8}S$, over a flat TiO_2 surface, demonstrating that the desired conformal film could be formed.

8.8 Anti-corrosion Layers

Methods of coating metals to prevent corrosion have been studied for many years, and are very important commercially. When a metal part is designed to be used in a harsh environment, the designer has two choices to make. He can either choose an exotic and expensive metal or alloy that will resist the environment, or he can select a more common metal and select a coating to protect it. Among the methods for coating metals are sputtering, evaporation, CVD, as well as many others. The problem found when applying any of these procedures is that coatings are never perfect, and pinhole defects are extremely common. A pinhole provides a starting place for local corrosion, which eventually affects the entire part surface.

As far as corrosion prevention in terms of pinholes or other point defects in applied coatings, ALD has the unique feature that it coats every aspect of a surface with perfect conformality. Therefore, if the part has point defects such as pits or protrusions, or if an initial coating has the same characteristics, then an ALD coating will overcoat these defects and prevent corrosion. Two recent papers [64–65] demonstrate the successful use of amorphous TiO_2 ALD films to protect stainless steel. In one instance the bare metal was coated [64]. In the second, a stainless steel part that had previously been coated with CrN was coated [65]. In both cases an improved corrosion performance was documented. For the second case, the authors speculate that the improved results were due to the ability of the ALD film to cover point defects in the first coating.

An interesting and relatively new commercial application for ALD is protection of silver from tarnishing. Anti-tarnishing treatment of silver by ALD can provide advantages in silver protection itself, but can also provide ways to develop the production and operations in the manufacturing process. One commercial silver jewelry manufacturer using ALD is Finland based Kalevala Jewelry.

The commercial solution for silver protection against tarnishing by ALD is based on encapsulation of the silver item by a single or dual oxide layer. In this application aluminum oxide (Al_2O_3) and titanium oxide (TiO_2) ALD films are used. The protective layer is made so that it will not affect the unique appearance of silver itself, important in jewelry. This means that the coating has to be optically transparent in the visible light region. ALD as a layer by layer coating method provides an accurate process enabling control of the optical behavior of single and multilayer films. The optical characteristics of thin films on silver may also be used to enhance it's artistic look, providing metallic colors for example.

Anti-tarnishing of silver jewelry by ALD demonstrates that ALD is moderately easy to implement in existing production where neither ALD nor any other vacuum coating technology has been used before.

Today's ALD reactors are user friendly and often tailored specifically to the application area. In silver jewelry protection, ALD has replaced conventional finishing methods such as lacquering and chemical passivation. Implementation of ALD in the production has allowed a reduction in the use of harmful chemicals and therefore has provided a clear development step in production. Conventional methods may also have disadvantages that emerge in the final product. Thickness control in lacquering is difficult and a thick enough lacquer layer providing sufficient protection against tarnishing changes the silver natural appearance showing a plastic-like surface. In the jewelry industry this is a marketing issue. Thinner lacquer layer do not retain the glossy appearance of silver during the logistic chain from production to end customer. Thin nanometer scale ALD layer will not provide permanent protection on wearable silver jewelry, but is sufficient to maintain appearance during the logistic chain [66–67].

Furthermore jewelry has a large variety of surface morphology and the objects usually appear in complex shapes. Also, the product collection is changing on a regular basis. ALD can be seen to be an ideal coating method for these type of structures since the film growth is based on surface reactions and it can be expected to form a conformal and pinhole free layer on the surface regardless of object shape or orientation. Finally, the coating reactor, when designed for coating multiple and variously-shaped articles, is flexible in production, as shown in Figure 8.11 [67].

Figure 8.11 Jewelry in a Beneq TFS 500 chamber (Photo courtesy and Copyright Kalevala Jewelry Oy *2012*).

8.9 Opportunities in Organic Electronics

In recent years it has become of interest to see if thin film devices traditionally made from inorganic materials could be fabricated from organic materials instead. With the invention and development of organic semiconductors many possibilities are being investigated. In addition to integrated circuits, a great deal of attention has been paid towards the development of flexible, plastic displays as well as organic LEDs (OLED) for lighting [68–69]. In fact such devices are now being used in a number of commercial products. The problem with their development has been that when exposed to the atmosphere, water vapor and oxygen diffuse into them and eventually affect their performance. After much development work it was determined that no single thin film (either an organic or an inorganic) could encapsulate these devices adequately. Therefore, subsequent work focused on alternate layers of organic and inorganic materials [70–71]. In one case, the inorganic layer was deposited by PECVD [70,] and in the other case it was deposited by sputtering [71]. Although these multilayer coatings are apparently functional, the process of depositing up to 12 pairs of organic/inorganic materials is quite complicated.

The reason that a single layer of an inorganic material has difficulty sealing a plastic surface is because such surfaces are fairly non-uniform at the nanometer level, and neither CVD or sputtering are capable of sealing such a surface completely. On the other hand, an ALD film may be able to, since it can coat with perfect conformality and leave no pinholes. Early studies described a single layer thermal ALD deposition of Al_2O_3 onto flexible plastic substrates, for use in OLEDs, at temperatures up to 175°C [72]. Water vapor diffusion rates of $1x10^{-3}$ g/m^2/day were measured for a 260 Å thick film. Although not as good as the $1x10^{-6}$ g/m^2/day needed for this application [73], it showed very good performance for a single inorganic layer. Subsequent studies have shown that with ALD even lower water vapor transmission rates (WVTRs) are possible. WVTR of less than $5x10^{-5}$ g/m^2/day (38°C and 85% RH) for 10 nm single Al_2O_3 deposited at 125°C on PET and PEN substrates has been measured by the Ca test [74]. Plasma assistance in ALD to create radicals provides a way to lower the substrate temperature that is favorable for polymer processing. Hence, in addition to thermal ALD several investigations with REALD have focused on coating on polymers. Studies have shown that REALD can be used to provide good gas barrier properties on polymer substrates and to coat plastic parts for their further enhancement. The deposition temperature could even be as low as room temperature [75–79]. REALD process development is important, if not essential, to implement ALD for industrial flexible organic electronic applications. This is particularly important with roll-to-roll ALD process development, where the plasma assistance to produce radicals can be used to increase the web speed and consequently process efficiency [80]. The roll-to-roll ALD development is discussed more in Chapter 1.3.3.2.

Even though the single ultrathin ALD layer has great potential in encapsulation, further development is needed to meet the requirements of organic devices. One disadvantage of using the well studied single Al_2O_3 ALD layer in future OLED devices is that it is prone to water corrosion [81]. Therefore further improvement of the encapsulation layer has focused onmultilayer structures, and so called nanolaminates [81–82]. In one case an ALD Al_2O_3 layer was combined with rapid ALD SiO_2 layers. Even in a bilayer structure, the SiO_2 layer effectively protected Al_2O_3 from corrosion. A multilayer structure of more than two Al_2O_3/SiO_2 bilayers led to deterioration of the water vapor barrier. This was attributed to the larger thickness and brittleness of the multilayer structure. This is a serious disadvantage especially in applications based on flexible substrates [81]. In one study, ultrathin Al_2O_3 ALD was combined with a CVD grown SiN layer in order to prevent the Al_2O_3 corrosion under water vapor exposure in the same way as the Al_2O_3/SiO_2 bilayers [74].

Regardless of its mechanical robustness, the nanolaminate structure consisting of alternative layers of Al_2O_3 and ZrO_2 was studied as an encapsulation of complete p-i-n OLED structure built on ITO coated glass. It was shown that ALD can be used without damaging the organic functional layer and that the nanolaminate structure performed better as an encapsulation layer than a single Al_2O_3 layer [82]. Instead of using an Al_2O_3 barrier, TiO_2 has been shown to have high barrier properties when deposited at low temperature. Furthermore, it has been suggested that material costs associated with high volume production of TiO_2 can be even lower than for Al_2O_3, which then makes it of even more interest to industry [80].

In addition to encapsulation of organic devices, other possibilities for ALD can be seen in transparent conductive oxides (TCO) for both organic photovolatics (OPVs) and OLEDs. Early studies demonstrated the possibility of ZnO for this purpose and also suggested that doping of ZnO would further improve its characteristics as a conductor [83]. Recently low temperature ZnO has also been studied and demonstrated to be an effective electron selective and hole blocking layer between ITO and the active material layer in flexible OPV device [84]. It has also been shown recently that high quality aluminum doped ZnO can be deposited at low temperatures, offering a replacement for ITO [85]. This together with the fact that the active material used in OPVs is sensitive to ITO may open commercial opportunities for ALD.

Deposition of nanometer thick coatings by ALD for other applications benefiting from the properties of organic/inorganic hybrid materials is under active investigation. For example, coating of textiles is being explored [86–89]. Other industrial sectors such as paper and paper converting, food packaging and product labeling may find interesting applications along with the development of ALD towards large scale ALD processes on flexible substrates. Here ALD oxides have been studied for example as barrier layers on fiber and polymer based packaging materials [90–91] and deacidification and UV protective layer on paper [92].

8.10 ALD Tool Manufacturers and Coating Providers

The great potential of ALD can be illustrated by considering the ALD tool manufacturers and coating service providers. In fact, in addition to the review presented earlier, future and in some cases already existing industrial applications can be seen or predicted from the portfolios of ALD application, equipment and service vendors. Many of them have recognized its potential for example in MEMS/NEMS and in a range of electronic applications such as DRAM, magnetic heads, high-k dielectrics and encapsulation of devices. The following list of ALD vendors does not necessarily include every vendor in the business.

AIXTRON provides ALD systems for semiconductor processing applications. Applications include dielectrics and electrodes for DRAM and eDRAM, high-k gate dielectrics, dielectrics and electrodes for MIM capacitor, Flash inter-poly dielectrics, STI gap-fill oxides and liners, dielectrics and liners for spacer and contact etch-stop films, films for thin film heads [93].

ALD Nanosolutions Inc. offers solutions for deposition on nanoparticles and nanomaterials based on fluidized bed ALD. Their applications include batteries and energy, MEMS/NEMS, catalysts and ultra barriers [94].

Applied Materials operates in the semiconductor industry and focuses in their ALD tools on high-k gate oxide (HfO_2) for transistors, tungsten nucleation layers and TaN barrier and TiN for DRAM applications. Development in transistors is looking at 22 nm and below in CMOS node technology [95].

ASM International operates in the semiconductor industry and provide tools for ALD Al_2O_3, TiN, high-k gate dielectrics (hafnium oxide, hafnium silicate), high-k capping layers for metal gate work function tuning, high-speed aluminum oxide for magnetic heads, passivation layers, high-k for MEMS, metal gate layers in CMOS and capacitor electrodes, low-k dielectric films, and oxides, nitrides and carbon for gate protection [96].

Beneq Oy offers ALD equipment from small batch research tools to large scale batch tools,to high volume and high throughput systems. Beneq has also a system for spatial ALD research for flexible substrates and true roll-to-roll ALD system on flexible webs up to 500 mm width for industrial and research usage. Applications that Beneq offers are barrier coatings for flexible and organic electronics, buffer layers for CIGS solar cells, surface passivation for c-Si solar cells, crack resistant coatings on glass, biocompatible ALD films, decorative coatings, optical coatings, coatings on particles and anti-tarnish coatings on silver. Beneq offers also coating service in ALD [97].

Kurt J. Lesker offers ALD system for basic research and complementary integrated deposition systems for R&D. ALD applications among the research are semiconductor, photovoltaics and OLED and organic electronics [98].

Lotus Applied Technology is company who develops roll-to-roll ALD technology for innovative applications on flexible substrates [99].

Oxford Instruments offers ALD systems for research and production. Applications for ALD are in nano-electronics, high-k gate oxides, storage capacitor dielectrics, high aspect ratio diffusion barriers for Cu interconnects, passivation layers for OLED's and polymers, passivation of c-Si solar cells, coatings for microfluidic and MEMS applications, coating of nanoporous structures, bio MEMS and fuel cells [100].

Picosun Oy offers ALD reactors for research, product development, pilot production and ALD production with batch and high volume configurations. Applications for ALD are in semiconductors, encapsulation for OLED lighting, surface passivation of c-Si solar cells, transparent conductive oxide and buffer layers for CIGS solar cells, electrodes, electrolytes and barriers for lithium batteries, corrosion protection, optics and displays, coatings on particles, decorative coatings and antitarnishing [101].

SENTECH has plasma and thermal ALD systems available in their product portfolio [102].

SoLayTec offers process development and high throughput ALD systems for c-Si solar cell passivation based on spatial ALD [103].

SVT Associates Inc. offers research ALD equipment with applications in high-k dielectrics, nanocoatings, surface modification layers, corrosion protection, moisture barriers, device encapsulation, MEMS and photonic crystals [104].

References

1. M. T. Bohr, R. S. Chan, T. Ghani, K. Mistry, *IEEE Spectrum*, 44(10), 29 (2007).
2. "Five-level MEMS Create True Systems", R. DeJule, *Semiconductor International*, April 2000.
3. "The Little Machines that are Making it Big", D. Bishop, P. Gammel, and C. R. Giles, *Physics Today*, October 2001.
4. "MEMS Diversify and Advance", P. Singer, Semiconductor International, October 2002.
5. "A New Wave of Microfluidic Devices", J. Ouellette, *The Industrial Physicist*, August/September 2003.
6. US Patent 6,942,318 September 13, 2005 ; *US Patent* 7,025,894 April 11, 2006.
7. *US Patent Application* 2007/0030313 February 8, 2007.
8. M. Sasaki, T. Yamaguchi, J.-H. Song, and K. Hane, J. Lightwave Technol. 21, 602 (2003).
9. Larry Hornbeck, "Digital Light Processing and MEMS: reflecting the digital display needs of the networked society," Micro-Optical Technologies for Measurement, Sensors, and Microsystems, *Proc. of SPIE*, Vol. 2783, pp. 2–12 (1996).
10. *US Patent* 6,794,119 September 21, 2004.
11. W. S. Yang and S. W. Kang, *Thin Solid Films* **500**, 231 (2006).

12. T. M. Mayer, J. W. Elam, S. M. George, P. G. Kotula and R. S. Goeke, *Appl. Phys. Lett.* **82**, 2883 (2003).
13. N. D. Hoivik, J. W. Elam, R. J. Linderman, V. M. Bright, S. M. George and Y. C. Lee, *Sensors and Actuators A*: Physical **103**, (2003).
14. M. Ritala, J. Niinistö, *ECS Trans.* 25 (8) 641–652, (2009).
15. *US Patent* 6,759,081 B2, July 6, 2004.
16. M. Mao, R. Bubber, T. Schneider, *ECS Transact.* 1, 37 (2006).
17. D. M. King, X. Liang, A. W. Weimer, *Powder Technol.* 221, 13 (2012).
18. http://www.aldnanosolutions.com, press release on November 28, 2011.
19. J. A. McCormick, B. L. Cloutier, A. W. Weimer, S. M. George, *J. Vac. Sci. Technol. A* 25(1), 67 (2007).
20. J. Zhao, Y. Wang, *J. Phys. Chem. C*, 116, 11867 (2012).
21. X. Liang, D. M. King, M. D. Groner, J. H. Blackson, J. D. Harris, S. M. George, A. W. Weimer, *J. Membrane Sci.* 322, 105 (2008).
22. Y. Zhou, D. M. King, X. Liang, J. Li, A. W. Weimer, Appl. Catal. B-Environ. 101, 54 (2010).
23. "Catalysis and Surface Science", G. Ertl and H-J. Freund, *Physics Today*, January 1999.
24. S. Haukka, E-L. Lakomaa, O. Jylhä, J. Vilhunen and S. Hornytzkyj, *Langmuir* **9**, 3497 (1993).
25. M. Linblad, L. P. Lindfors and T. Suntola, *Catal. Lett.* **27**, 323 (1994).
26. E-L. Lakomaa, *Appl. Surf. Sci* **75**, 185 (1994).
27. M. Linblad, S. Haukka, A. Kytökivi, E-l. Lakomaa, A. Rautianen and T. Suntola, *Appl. Surf. Sci.* **122/122**, 286 (1997).
28. N-T. Nguyen and S. H. Chan, J. *Micromech. Microeng.*, **16**, R1 (2006).
29. A. Weber and E. Ivers-Tiffee, *J. Power Sources* **127**, 273 (2004).
30. C. Bernay, A. Ringuedé, P. Colomban, D. Lincot, M. Cassir, *J. Phys. Chem. Solids* 64, 1761 (2003).
31. C. N. Ginestra, R. Sreenivasan, A. Karthikeyan, S. Ramanathan and P. C. McIntyre, *Electrochem. Solid-State Lett.*, **10**, B161 (2007).
32. X. Jiang, J. S. King and S. F. Bent, "Atomic Layer Deposition of Alternative Energy Materials", American Vacuum Society 54th International Symposium, Seattle, WA (2007).
33. B. Medina-Lott, M. Tasse, C. Brahim, A. Ringuede, L. Niinistö, M. Cassir, *ECS Transact.* 35(1), 601 (2011).
34. J. W. Elam, N. P. Dasgupta, F. B. Prinz (2011). ALD for clean energy conversion, utilization, and storage. *MRS Bulletin*, 36, 899–906.
35. "Membranes for Hydrogen Purification", *MRS Bulletin* **31**, October 2006.
36. M. A. Cameron, J. P. Gartland, J. A. Smith, S. F. Diaz and S. M. George, *Langmuir* **16**, 7435 (2000).
37. S. Zaitsu, S. Motokoshi, T. Jitsuno, M. Nakatsuka and T. Yamanaka, *Jpn. J. Appl. Phys.* **41**, 160 (2002).

38. G. Stuyven, P. De Visschere, K. Neyts and A. Hikavvy, *Jpn. J. Appl. Phys.* **41**, 5702 (2002).
39. http://www.apitech.com, press release Nov. 8th 2007 (June 27th 2012).
40. http://www.apitech.com, press release Jan. 15th 2008 (June 27th 2012).
41. T. Suntola, J. Antson, *US Patent* 4058430 (1997).
42. T. Suntola and M. Simpson (eds.), *Atomic Layer Epitaxy*, Blackie and Sons Ltd, London, 1990.
43. A. Pakkala, "EL Displays Based on ALE Grown Phosphors and Insulator Thin Films", white paper, Planar International Ltd. http://www.planar.com/support/design- resources/docs/alephosph.pdf, accessed 20.3.2012.
44. "Solar Energy Conversion", G. W. Crabtree and N. S. Lewis, *Physics Today*, March 2007.
45. B. Sang, A. Yamada and M. Konagai, *Jpn. J. Appl. Phys.* **37**, L206 (1998).
46. S. Chaisitsak, T. Sugiyama, A. Yamada and M. Konagai, *Jpn. J. Appl. Phys.* **38**, 4989 (1999).
47. C. Platzer-Björkman, A. Hultqvist, J. Pettersson, T. Törndahl, *Proc. SPIE* 7603, 76030F (2012).
48. C. Platzer-Björkman, T. Törndahl, D. Abou-Ras, J. Malmström, J. Kessler, L. Stolt, *J. Appl. Phys.* 100, 044506 (2006).
49. C. Platzer-Björman, T. Törndahl, A. Hultqvist, J. Kessler, M. Edoff, *Thin Solid Films*, 515, 6024 (2007).
50. A. Hultqvist, C. Platzer-Björkman, E. Coronel, M. Edoff, *Sol. Energy Mater. Sol. Cells* 95, 497 (2011).
51. http://www.beneq.com/buffer-layer-cigs.html (April 30th 2012).
52. H. Lee, T. Tanchibana, N. Ikero, H. Hashiguchi, K. Arafune, H. Yoshida, S-I. Satoh, T. Chikyow, A. Ogura, *Appl. Phys. Lett.* 100, 143901 (2012).
53. T. Lüder, G. Hahn, B. Terheiden, *Energy Procedia* 8 660 (2011).
54. J. Schmidt, A. Merkle, R. Brendel, B. Hoex, M. C. M. van de Sanden, W. M. M. Kessels, *Prog. Photovolt: Res. Appl.* 16, 461 (2008).
55. B. Hoex, J. Schmidt, R. Bock, P. P. Altermatt. M. C. M. van de Sanden, W. M. M. Kessels, *Appl. Phys. Lett.* 91, 112107 (2007).
56. G. Dingemans, R. Seguin, P. Engelhart, M. C. M. van de Sanden, W. M. M. Kessels, *Phys. Status Solidi RRL* 4, 10 (2010).
57. J. I. Skarp, *ECS Trans.* 33 447 (2010).
58. V. I. Kuznetsov, E. H. A. Granneman, P. Vermont, K. Vanormelingen, *ECS Trans.* 33, 441 (2010).
59. P. Poodt, V. Tiba, F. Werner, J. Schmidt, A. Vermeer, F. Roozeboom, *J. Electrochem. Soc.* 158(9), H937 (2011).
60. F. Werner, W. Stals, R. Görtzen, B. Veith, R. Brendel, J. Schmidt, *Energy Procedia* 8, 301 (2011).
61. http://www.beneq.com/surface-passivation-c-si.html (May 20th 2012).

62. http://www.solaytec.com/products/high-volume-tool-2100/3600 (June 14th 2012).
63. L. Reijnen, B. Meester, A. Goossens and J. Schoonman, *Chem. Vap. Deposition* **9,** 15 (2003).
64. C. X. Shan, X. Hou and K-L. Choy, *Surf. Coat. Technol.* 202, 2399 (2008).
65. C. X. Shan, X. Hou and K-L. Choy, *Surf. Coat. Technol.* 202, 2147 (2008).
66. M. Elamo, M. Laakso, Interview of Kalevala Jewelry representatives, September 2011.
67. http://www.beneq.com/silver-anti-tarnish-coating.html (January 31st 2012).
68. *U.S. Patent* 6,866,949 March 15, 2005.
69. W. A. MacDonald, *J. Mater. Chem.* **14,** 4 (2004).
70. T. W. Kim, M. Yan, A. G. Erlat, P. A. McConnelee, M. Pellow, J. DeLuca, T. P. Feist and A. R. Duggal, *J. Vac. Sci. Technol.* **A23**, 971 (2005).
71. L. Moro, T. A. Krajewski, N. M. Rutherford, O. Philips, R. J. Visser, M. Gross, W. D. Bennett and G. Graff, Organic Light-Emitting Materials and Devices VII, edited by Z. H. Kafafi and P. A. Lane, *Proceedings of SPIE* **5214**, 83 (2004).
72. M. D. Groner, S. M. George, R. S. Mclean and P. F. Garcia, *Appl. Phys. Lett.* **88**, 051907 (2006).
73. J. Lewis. *Materials Today*, Vol 9, 2004.
74. P. F. Carcia, R. S. Mclean, M. D. Groner, A. A. Dameron, S. M. George, *J. Appl. Phys.* 106, 023533, (2009).
75. S. J. Yun, Y-W. Koo and J. W. Lim, *Appl. Phys. Lett.* **85**, 4896 (2004).
76. S. J. Yun, J. W. Lim and J-H. Lee, *Electrochem. Solid-State Lett.***7**, C13 (2004).
77. E. Langereis, M. Creatore, S. B. S. Heil, M. C. M. van de Sanden and W. M. M. Kessels, *Appl. Phys. Lett.* **89**, 081915 (2006).
78. T. O. Kääriäinen, D. C. Cameron, *Plasma Process. Polym.* 6, S237 (2009).
79. W-S. Kim, D-Y. Moon, B-W. Kang, J-W. Park, *Korean Phys. Soc.* 55, 55 (2009).
80. E. Dickey, W. A. Barrow, *J. Vac. Sci. Technol. A* 30(2), 021502-2 (2012).
81. A. A. Dameron, S. D. Davidson, B. B. Burton, P. F. Carcia, R. S. McLean, S. M. George, *J. Phys. Chem. C.* 112, 4573 (2008).
82. J. Meyer, D. Schneidenbach, T. Winkler, S. Hamwi, T. Weimann, P. Hinze, S. Ammermann, H.-H. Johannes, T. Riedl, W. Kowalsky, *Appl. Phys. Lett.* 94, 233305 (2009).
83. A. W. Ott, R. P. H. Chang, *Mat. Chem. Phys.*58, 132 (1999).
84. J-C. Wang, W-T. Weng, M-Y. Tsai, M-K. Lee, S-F. Horng, T-P. Perng, C-C. Kei, C-C. Yu, H-F. Meng, *J. Mater. Chem.* 20, 862 (2010).
85. H. Saarenpää, T. Niemi, A. Tukiainen, H. Lemmetyinen, N. Tkachenko, *Sol.Energy Mater. Sol. C.* 94, 1379 (2010).
86. G. K. Hyde, K. J. Park, S. M. Stewart, J. P. Hinestroza, G. N. Parsons, *Langmuir* 23, 9844 (2007).

87. *US Patent* 20090137043, May 28, 2009.
88. K. G. Hyde, G. Scarel, J. C. Spagnola, Q. Peng, K. Lee, B. Gong, K. G. Roberts, K. M. Roth, C. A. Hanson, C. K. Devine, S. M. Stewart, D. Hojo, J-S. Na, J. S. Jur, G. N. Parsons, *Langmuir* 26(4), 2550 (2010).
89. J. S. Jur, W. J. Sweet III, C. J. Oldham, G. N. Parsons, Adv. Funct. Mater. 21, 1993 (2011).
90. T. Hirvikorpi, M. Vähä-Nissi, T. Mustonen, E. Iiskola, M. Karppinen, *Thin Solid Films* 518, 2654 (2010).
91. T. O. Kääriäinen, P. Maydannik, D. C. Cameron, P. Johansson, K. Lahtinen, J. Kuusipalo. *Thin Solid Films* 519, 3146 (2011).
92. C. A. Hanson, C. J. Oldham, G. N. Parsons, *J. Vac. Sci. Technol. A.* 30(1), 01A117/1 (2012).
93. http://www.aixtron.com (June 27th 2012).
94. http://www.aldnanosolutions.com (June 27th 2012).
95. http://www.appliedmaterials.com (June 27th 2012).
96. http://asm.com (June 27th 2012).
97. http://www.beneq.com (June 27th 2012).
98. http://lesker.com (June 27th 2012).
99. http://www.lotusat.com (June 27th 2012).
100. http://www.oxford-instruments.com (June 27th 2012).
101. http://www.picosun.com (June 27th 2012).
102. http://www.sentech.com (June 27th 2012).
103. http://www.solaytec.com (June 27th 2012).
104. http://www.svta.com (June 27th 2012).

Index

Also of Interest

Check out these published and forthcoming related titles from Scrivener Publishing

Advances in Contact Angle, Wettability and Adhesion Volume 1
Edited by K.L. Mittal
June 2013. ISBN 978-1-118-47292-7

Advances in Modeling and Design of Adhesively Bonded Systems
Edited by S. Kumar and K.L. Mittal
May 2013. ISBN 978-1-118-68637-9

Atomic Layer Deposition Principles, Characteristics, and Nanotechnology Applications
By Tommi Kääriäinen, David Cameron, Marja-Leena Kääriäinen and Arthur Sherman
Published 2013. ISBN 978-1-118-06277-7

Atmospheric Pressure Plasma Treatment of Polymers
Edited by Michael Thomas and K.L. Mittal
Published 2013. ISBN 978-1-118-59621-0

Encapsulation Nanotechnologies
Edited by Vikas Mittal
Published 2013. ISBN 978-1-118-34455-2

Atmospheric Pressure Plasma for Surface Modification
By Rory A. Wolf
Published 2012. ISBN 9781118016237

Introduction to Surface Engineering and Functionally Engineered Materials
By Peter Martin
Published 2011. ISBN 978-0-470-63927-6

Plastics Sustainability
Towards a Peaceful Coexistence between Bio-based and Fossil fuel-based Plastics
By Michael Tolinski
Published 2011. ISBN 978-0-470-93878-2

Roll-to-Roll-Vacuum Deposition of Barrier Coatings
By Charles A. Bishop
Published 2010. ISBN 978-0470-60956-9

Miniemulsion Polymerization Technology
edited by Vikas Mittal
Published 2010. ISBN 978-0-470-62596-5

www.ingramcontent.com/pod-product-compliance
Lightning Source LLC
La Vergne TN
LVHW011753200926
848484LV00003B/695